What Every Engineer Should Know About Data-Driven Analytics

What Every Engineer Should Know About Data-Driven Analytics provides a comprehensive introduction to the theoretical concepts and approaches of machine learning that are used in predictive data analytics. By introducing the theory and providing practical applications, this text can be understood by students of every engineering discipline. It offers a detailed and focused treatment of the important machine learning approaches and concepts that can be exploited to build models to enable decision making in different domains.

- Utilizes practical examples from different disciplines and sectors within engineering and other related technical areas to demonstrate how to go from data, to insight, and to decision making.
- Introduces various approaches to building models that exploit different algorithms.
- Discusses predictive models that can be built through machine learning and used to mine patterns from large datasets.
- Explores the augmentation of technical and mathematical materials with explanatory worked examples.
- Includes a glossary, self-assessments, and worked-out practice exercises.

Written to be accessible to non-experts in the subject, this comprehensive introductory text is suitable for students, professionals, and researchers in engineering and data science.

What Every Engineer Should Know
Series Editor
Phillip A. Laplante
Pennsylvania State University

What Every Engineer Should Know about MATLAB® and Simulink ®
Adrian B. Biran

Green Entrepreneur Handbook: The Guide to Building and Growing a Green and Clean Business
Eric Koester

What Every Engineer Should Know about Cyber Security and Digital Forensics
Joanna F. DeFranco

What Every Engineer Should Know about Modeling and Simulation
Raymond J. Madachy and Daniel Houston

What Every Engineer Should Know about Excel, Second Edition
J.P. Holman and Blake K. Holman

Technical Writing: A Practical Guide for Engineers, Scientists, and Nontechnical Professionals, Second Edition
Phillip A. Laplante

What Every Engineer Should Know about the Internet of Things
Joanna F. DeFranco and Mohamad Kassab

What Every Engineer Should Know about Software Engineering
Phillip A. Laplante and Mohamad Kassab

What Every Engineer Should Cyber Security and Digital Forensics
Joanna F. DeFranco and Bob Maley

Ethical Engineering: A Practical Guide with Case Studies
Eugene Schlossberger

What Every Engineer Should Know About Data-Driven Analytics
Satish Mahadevan Srinivasan and Phillip A. Laplante

What Every Engineer Should Know About Reliability and Risk Analysis
Mohammad Modarres and Katrina Groth

For more information about this se ries, please visit: www.routledge.com/What-Every-Engineer-Should-Know/book-series/CRCWEESK

What Every Engineer Should Know About Data-Driven Analytics

Satish Mahadevan Srinivasan
Phillip A. Laplante

CRC Press is an imprint of the
Taylor & Francis Group, an **informa** business

Cover image: Shutterstock

First edition published 2023
by CRC Press
6000 Broken Sound Parkway NW, Suite 300, Boca Raton, FL 33487-2742

and by CRC Press
4 Park Square, Milton Park, Abingdon, Oxon, OX14 4RN

CRC Press is an imprint of Taylor & Francis Group, LLC

© 2023 Phillip A. Laplante and Satish Mahadevan Srinivasan

Reasonable efforts have been made to publish reliable data and information, but the author and publisher cannot assume responsibility for the validity of all materials or the consequences of their use. The authors and publishers have attempted to trace the copyright holders of all material reproduced in this publication and apologize to copyright holders if permission to publish in this form has not been obtained. If any copyright material has not been acknowledged please write and let us know so we may rectify in any future reprint.

Except as permitted under U.S. Copyright Law, no part of this book may be reprinted, reproduced, transmitted, or utilized in any form by any electronic, mechanical, or other means, now known or hereafter invented, including photocopying, microfilming, and recording, or in any information storage or retrieval system, without written permission from the publishers.

For permission to photocopy or use material electronically from this work, access www.copyright.com or contact the Copyright Clearance Center, Inc. (CCC), 222 Rosewood Drive, Danvers, MA 01923, 978-750-8400. For works that are not available on CCC please contact mpkbookspermissions@tandf.co.uk

Trademark notice: Product or corporate names may be trademarks or registered trademarks and are used only for identification and explanation without intent to infringe.

ISBN: 978-1-032-23543-1 (hbk)
ISBN: 978-1-032-23540-0 (pbk)
ISBN: 978-1-003-27817-7 (ebk)

DOI: 10.1201/9781003278177

Typeset in Times
by SPi Technologies India Pvt Ltd (Straive)

Dedication

Each author would like to thank his respective family members, parents, grandparents, great-grandparents, and so on down the line. Without these ancestors the authors and this book would never exist.

This book is dedicated to the memory of our dear colleague and gentle friend Partha Mukherjee who sadly passed away before this book was completed.

Contents

Preface

INTRODUCTION

This book provides a comprehensive introduction to the machine learning theoretical concepts and approaches that are used in predictive data analytics through practical applications (case studies and examples). Using machine learning we can build predictive models that can be used to mine patterns from large datasets. Such models can also be tailored to reason why it sees a particular pattern in the dataset.

Mining large datasets from different domains (healthcare, financial, sports, manufacturing, social media, advertisement, etc.) needs a different type of mindset and skillset for predictive model building. This textbook will offer a detailed and focused treatment of the important machine learning approaches and concepts that can be exploited to build models to enable decision making in different domains. Whenever required, technical and mathematical materials will be augmented with explanatory worked examples to illustrate their importance in the given context.

Through case studies, this book demonstrates how to go from data, to insight, to decision making. In each of the case studies, we have taken a unique approach to building models that exploit different algorithms. In addition to that, the case studies also highlight the techniques used for validating and evaluating predictive models. The book, informed by the author's many years of teaching machine learning, and working on predictive data analytics projects, is suitable for use by graduates, professionals, and researchers in the area of data science.

AUDIENCE

This book is intended for professional data engineers, software engineers, systems engineers, and senior and graduate students of analytics and artificial intelligence. Much of the material is derived from the graduate-level "Data Analytics" course taught at Penn State's Great Valley School of Graduate and Professional Studies and online through its World Campus, where the authors work. The typical student in that course has five years of work experience in any of a variety of technical or business roles and an undergraduate degree in engineering, science, or business. Typical readers of this book will have one of the following or similar job titles:

Data analyst
Data scientist
IT analyst
Software engineer
Systems engineer
Sales engineer
Systems analyst
[XYZ] engineer (where "XYZ" is an adjective for most engineering disciplines, such as "electrical," "computer," or "mechanical")

Project manager
Business analyst
Technical architect
Lead architect
Product owner

Many others can benefit from this text including the users of complex systems and other stakeholders.

COURSE ADOPTION

This text is suitable for use in the following courses as a primary reference: predictive analytics, machine learning, data-driven decision making, and data science.

It can also be used as a secondary reference, typically in courses such as data mining, statistics, natural language processing, and artificial intelligence.

ERRORS

The authors have tried to uphold the highest standards for accuracy in terms of fact and quality of presentation. Despite these best efforts and those of the reviewers and publisher, there are still likely a few errors to be found. Therefore, if you believe that you have found an error—whether it is a referencing issue, factual error, or typographical error—please contact the authors at sus64@psu.edu or pal11@psu.edu.

Acknowledgments

There are many who helped in the development and writing of this book, directly or indirectly. In particular, we would like to express our gratitude to our Senior Publisher and Sponsor at T&F, Allison Shatkin, for her support and constant encouragement along the way. We would also like to thank the following students for assisting with some of the code examples, figures, and tables.

Shahed Mahbub, Raghava Rao Sunkanapally, Nikhitha Kunduru, Junjun Tao, Shichu Chen, Deeksha Joshi, Haruka George, Dinesh Kumar Chowdhary and Manish Ranjan.

We would also like to thank the following colleagues for their encouragement and ideas, and for reviewing portions of the draft manuscript.

Raghvinder S. Sangwan, Youakim Badr, Partha Mukherjee

There are surely others who have provided us with additional inspiration along the way and the omission of their names is inadvertent, but we would like to thank them all collectively.

About the Authors

Satish Mahadevan Srinivasan (Satish Srinivasan) is an Associate Professor of Information Science and a member of the graduate faculty at the Pennsylvania State University. His research, teaching, and consulting focus on predictive analytics particularly with respect to text preprocessing, building data pipelines, and artificial intelligence systems.

Satish Srinivasan received his B.E. in Information Technology from Bharathidasan University, India, and his M.S. in Industrial Engineering and Management from the Indian Institute of Technology Kharagpur, India. He earned his Ph.D. in Information Technology from the University of Nebraska at Omaha. Prior to joining Penn State Great Valley, he worked as a postdoctoral research associate at the University of Nebraska Medical Center, Omaha. Dr. Srinivasan teaches courses related to database design, data mining, data collection and cleaning, computer, network and web securities, and business process management. He has authored or edited 2 book chapters and more than 50 papers, articles, reviews, and editorials.

Phillip A. Laplante is Professor of Software and Systems Engineering and a member of the graduate faculty at the Pennsylvania State University. His research, teaching, and consulting focus on software quality, particularly with respect to artificial intelligence and critical systems.

Prior to his academic career, Dr. Laplante spent nearly a decade as a software engineer and project manager working on avionics (including the Space Shuttle), CAD, and software test systems. He has authored or edited 39 books and more than 300 papers, articles, reviews, and editorials.

Dr. Laplante received his B.S., M.Eng., and Ph.D. in computer science, electrical engineering, and computer science, respectively, from the Stevens Institute of Technology and an M.B.A. from the University of Colorado at Colorado Springs. He is a Licensed Professional Engineer in Pennsylvania and is a Certified Software Development Professional. He is a fellow of the IEEE and SPIE and a member of numerous professional societies and program committees.

1 Data Collection and Cleaning

In the 21st century, data are everywhere. Across different application areas, data are being collected at an unprecedented rate. Decisions in past were purely made based on guesswork, expert opinions, or by using constructed models; but these days decisions are made solely based on the data available. Large amounts of data or so-called "Big Data" has the potential to revolutionize different aspects of modern society. Application areas of Big Data include scientific research, financial services, retail manufacturing, biological and physical sciences, healthcare, transportation, environmental modeling, energy saving, homeland security, social network analysis, and much more [1].

So how should an engineer or data scientist approach the analysis of all this data? Here is the general approach. After recording the data in the repositories, the next step is to curate and analyze the data. The discussions in this book will focus on analyzing the data using the tools and techniques in the domain of machine learning, data mining, and predictive analytics. The potential uses of Big Data are exciting. For example, in the education sector, the collected data related to the academic performance of every student can be used as a guide for delivering future instructions. In the healthcare sector, Information Technology (IT) and data analytics can reduce the cost of healthcare while improving its quality and outcomes by making preventive care more personalized and affordable. In the United States alone, the advent of Big Data technology can result in IT savings for the healthcare sector, which is estimated to be close to 300 billion dollars [1, 2].

While the potential benefits are significant, there remain many technical challenges to be addressed for realizing the potential of Big Data. Challenges exist along several different dimensions, namely: *Volume*, *Variety*, *Velocity*, *Veracity*, and *Value* [1–3].

The dimension of *volume* represents the amount of data. The collected data can be either structured (numeric, relational model) or unstructured (non-numeric, text type, video, audio) which is represented by the dimension of *variety*. The *velocity* dimension represents the rate at which the data arrive over time (every second, every minute, hourly, daily, etc.) and also accounts for the time within which the data have to be acted upon. The dimension of *veracity* is all about the validity and the correctness of data, i.e., how accurate and usable are the data? Finally, how valuable are the data captured by the *value* dimension? [1–3]

A data analysis pipeline to deal with the different dimensions of Big Data is shown in Figure 1.1. The data analysis pipeline includes multiple phases, namely the *data acquisition/recording phase*, the *extraction/cleaning/annotation phase*, the *integration/aggregation/representation phase*, the *analysis/modeling phase*, and the

DOI: 10.1201/9781003278177-1

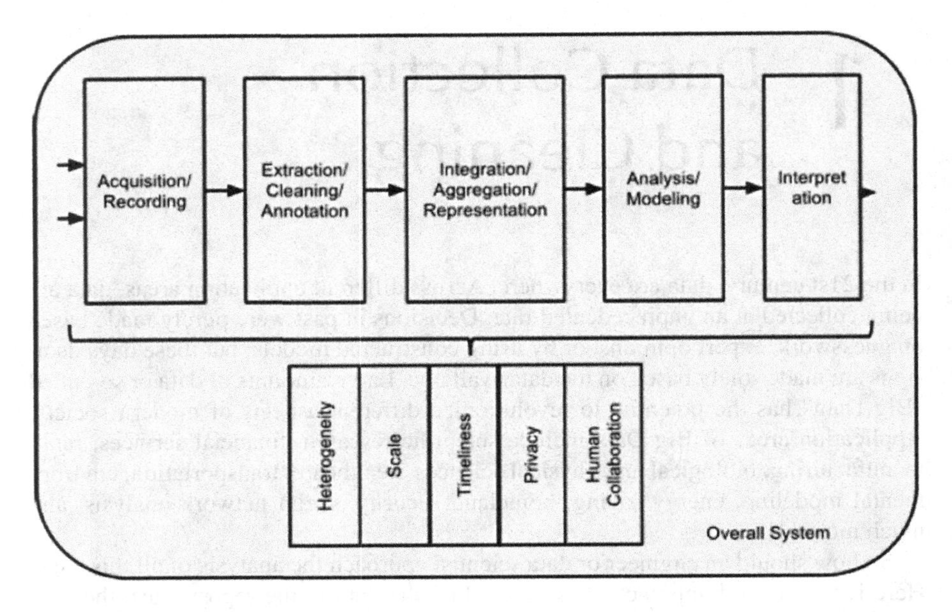

FIGURE 1.1 The Big Data analysis pipeline. (Source: Figure adapted from Challenges and Opportunities with Big Data, A community white paper developed by leading researchers across the United States, https://cra.org/ccc/wp-content/uploads/sites/2/2015/05/bigdata whitepaper.pdf.)

interpretation phase. Each of these phases is crucial and has its own set of challenges that needs to be addressed [1].

In the *data acquisition phase*, the challenge is to generate the right metadata to describe what data needs to be recorded and measured. *Data provenance*[1] is also an issue in this phase. Any data not originating from its source would have gone through several stages of transformation or edition. Therefore, an error in the data processing can render subsequent analysis useless. Thus, in the data analysis pipeline, it is important to carry both the provenance of data and metadata together [1].

The *information extraction and cleaning phase* is responsible for converting the collected data in a format that is ready for analysis [1].

The *data integration, aggregation, and representation phase* is devoted to hiding the heterogeneity of the data and making it available in the required format for analysis and modeling [1].

The *query processing, data modeling, and analysis phase* is devoted to building tools and techniques for the effective large-scale analysis of data in a completely automated manner [1].

Finally, the *interpretation phase* provides the means for the decision makers to interpret the results of the analysis and make Big Data more actionable [1].

Now let us focus our discussion on the challenges associated with the data analysis pipeline. The challenges involved in the analysis pipeline can be classified as *heterogeneity*, *scale*, *timelines*, *privacy*, and *human collaboration* [1].

The *heterogeneity and incompleteness* are challenges because the data representing the same entity in different sources lack a consistent format, are erroneous, and incomplete. Again, managing large and rapidly increasing volumes of data challenges the limitation of the tools, techniques, and the algorithms that process them. Often analyzing the large datasets takes a longer time, which is a challenge because with time, the value of the data diminishes for a decision maker [1].

Privacy in Big Data is a major concern as still there is no established protocol that allows the sharing of private data while limiting the disclosure and ensuring sufficient data utility [1].

Finally, the advances in computing analysis have still left a gap in identifying many patterns that are only detectable by humans. This is a challenge to the automation of the data analysis pipeline as *human intervention and collaboration* are of paramount importance in the successful realization of the potentiality of the data [1].

DATA-COLLECTION STRATEGIES

Big Data collection is the methodical approach to collecting and analyzing massive amounts of information from a variety of sources. Earlier it was indicated that Big Data collection entails collecting structured, semi-structured, and unstructured data generated by sources such as people, computers, and sensors. The value of the data does not depend on its quantity but on its quality. Structured data are highly organized and exist in a predefined format. On the other hand, there is no predefined format for unstructured data. Therefore, it exists in the format in which it was generated. Semi-structured data on the other hand is a mix of structured and unstructured data. For example, data related to GPS[2] coordinates is an example of structured data. Data collected from social media sites is a good example of unstructured data. Data like email addresses and their contents are good examples of semi-structured data. Data can also be classified as quantitative and qualitative. Quantitative data have numerical forms such as statistics and percentages, while qualitative data are more descriptive in nature, like sex, religion, etc. The typical sources of data include [1–3]:

- Operational systems producing transactional data,
- IoT endpoint device,
- Social media data from users and customers,
- Location data and other vitals from the smartphone devices,
- and more.

To provide access to accurate and consistent data, integration of data from several sources into a *data warehouse*[3] is very vital. Data warehouses require and provide extensive support for data cleaning. Since data warehouses need to load and continuously refresh huge amounts of data from a variety of sources, the probability of data being *dirty*[4] or inconsistent is very high. In order to provide effective and timely responses to queries, data cleaning in data warehouses is important and is supported by a so-called *ETL* process. The *ETL* process comprises of three phases namely the *extraction phase*, the *transformation phase*, and the *loading phase* [1, 3].

During the *extraction phase* data from several sources are collected. The source systems files (data) can be in multiple file formats including flat files with delimiters (CSV format), XML, non-relational database structures including IMS (Information Management Systems), data structures such as VSAM (Virtual Storage Access Method) or ISAM (Indexed Sequential Access Method) or data fetched through screen-scraping or web spidering. A source for data could also be a legacy system in which the files are in the arcane format. In this step, the schema (metadata) from different sources is extracted and translated. In addition to the schemas, instances are also extracted from the data source and are moved/stored into intermediate data sources before loading them into the data staging area [1, 3].

In the *transformation stage*, multiple data manipulation steps are performed such as moving, splitting, translating, merging, sorting, and pivoting data, all in accordance with the data quality rules. In this phase, the translated schema is matched and integrated to create a unique schema for the data warehouse. In addition to the schemas, the instances are also matched and integrated into the data staging area. A series of rules or functions is applied on the data extracted from the data sources which includes selective loading of the columns, translation of the coded values, encoding of free-form values, deriving calculated values, sorting, joining data from multiple sources, aggregation, transposing, etc. A large number of tools of varying functionalities are available to support these tasks, but often a significant portion of the cleaning and transformation work has to be done manually or by a low-level program that is difficult to write and maintain [1, 3].

Finally, in the *loading stage*, the translated and integrated schema from the transformation is implemented on the target (data warehouse) system and the instances from the data staging area are filtered, aggregated, and loaded into the data warehouse. Depending on the organization requirements, the process of loading the extracted data into the data warehouse is frequently done on a daily, weekly, or monthly basis [1, 3].

DATA PREPROCESSING STRATEGIES

Data preprocessing is performed to transform the raw data into a useful and efficient format. Major tasks involved in the data preprocessing stage are data cleaning, data integration, data reduction, data transformation, and data discretization. Data collection results in accumulating noisy, missing, and inconsistent data which need to be corrected for downstream analysis. Data preprocessing avoids any potential problems with accuracy, completeness, consistency, timeliness, believability, and interpretability [1–5].

Data Cleaning: Data can have many irrelevant and missing parts. Therefore, it is important to perform *data cleaning* to deal with missing and noisy data. Missing data can be handled by either ignoring the entire tuple or by imputing the missing values. On the other hand, noisy data can be handled by using the binning method, regression, or clustering [1, 3–5].

Data Integration: In this step tasks such as entity identification, removal of redundant data, and data deduplication are performed. Data fused from different sources and single entities from two different sources can have attributes that are referred to different naming conventions or might be referring to the same characteristic. Redundant attributes can be detected by correlation and covariance analysis. Duplicated data can be identified by recognizing the repeated tuples or by performing descriptive statistics [1, 3–5].

Data Reduction: Data reduction is the process to obtain a reduced representation of the dataset that is much smaller in volume but yet produces the same analytical results. A database or a data warehouse may store terabytes of data and analyzing this amount of data can take a very long time. Therefore, it is important to perform data reduction. Data reduction strategies include dimensionality reduction (attribute subset selection, attribute creation, wavelet transformation, principal components analysis (PCA), etc.), numerosity reduction (regression and log-linear models, histograms, clustering, sampling, data cube aggregation, etc.), data compression (string compression, audio/video compression, etc.), etc [1, 3–5].

Data Transformation: Data transformation is the process of converting data from one type to another. The strategy here is to map the values of a given attribute to a new set. The motivation behind the data transformation is to remove the skewness in the data to achieve symmetric distribution. Transforming data makes it easier to visualize, and to improve the data interpretability. Data transformations involve different operations such as smoothing to remove noise from the data, attribute/feature construction, normalization i.e., to scale data to fall within a smaller and specified range, discretization where raw values are replaced by intervals or conceptual labels, min-max normalization, z-score normalization, normalization by decimal scaling, concept hierarchy generation, etc [1, 3–5].

Data Discretization: Data discretization is a process of converting a large number of data values into smaller ones. Data discretization is performed in order to make data evaluation, visualization, and management easier [1, 3–5].

A brief introduction to the basics of R and Python programming is provided here, which will be very helpful for the readers to navigate through the other chapters in this book [2].

PROGRAMMING WITH R

DATA TYPES IN R

The basic data types in R are character, numeric, integer, and logical. The assignment operator (←) can be used to assign any data to a variable.

```
For example
x <- 10
y <- 'This is a sample string'
```

```
# Let us now print the value of x & y
x
[1] 10
y
[1] "This is a sample string"
```

Data Structures in R

The commonly used data structures in R include vectors, matrices, dataframe, lists, and factors.

Vectors—A vector is a sequence of data elements of the same basic type. A vector can be defined as

```
a <- c(1,2,5.3,6,-2,4) # numeric vector
b <- c("one", "two", "three") # character vector
c <- c(TRUE, TRUE, TRUE, FALSE, TRUE, FALSE) #logical vector
```

Matrices—A matrix is a collection of data elements arranged in a two-dimensional rectangular layout. For example, the matrix A of size 2 × 3 can be created as:

```
A = matrix(
c(2, 4, 3, 1, 5, 7), # the data elements
nrow=2,              # number of rows
ncol=3,              # number of columns
byrow = TRUE)        # fill matrix by rows
```

The matrix A can then be viewed by just entering the matrix name

```
A
     [,1] [,2] [,3]
[1,]   2    4    3
[2,]   1    5    7
```

The elements of the matrix can be directly accessed by specifying the corresponding row and column number as shown

```
A[2, 3]        # element at 2nd row, 3rd column
[1] 7
```

Dataframe—Compared to matrices, in a dataframe, the different columns can have different modes (numeric, character, factor, etc.). A dataframe in R can be created as

```
d <- c(1,2,3,4)
e <- c("green", "blue", "green", NA)
f <- c(TRUE,TRUE,TRUE,FALSE)
```

```
mydata <- data.frame(d,e,f) # create a data frame called
   mydata
names(mydata) <- c("ID","Color","Passed") # variable names
```

Now let us display the created dataframe *mydata*

```
mydata
   ID Color Passed
1  1  green  TRUE
2  2  blue   TRUE
3  3  green  TRUE
4  4  <NA>   FALSE
```

Lists—A list is defined as an ordered collection of objects. The objects can be of different types and possibly unrelated. For example,

```
# Example of a list with 4 components -
# a string, a numeric vector, a matrix, and a scaler
w <- list(name="Sam", mynumbers=xyz, mymatrix=abc, age=5.3)
# Example of a list containing two lists
v <- c(list1,list2) # where list1 and list2 are lists
```

Factors—Conceptually speaking factors are variables in R which take on a limited number of different values. These variables are often referred to as categorical variables.

```
# Define the variable gender with 20 "male" entries and 30
   "female" entries
gender <- c(rep("male",20), rep("female", 30))
gender <- factor(gender)
# The gender variable stores gender as 20 1s and 30 2s and
   associates 1 to male and 2 to female

# R now treats gender as a nominal variable
summary(gender)
female   male
  30      20
```

An ordered factor is used to represent an **ordinal variable**.

```
# A variable rating can be coded as "large", "medium", "small'
rating <- c("large", "medium", "small")
rating <- ordered(rating)
# recodes the variable rating to 1,2,3 and associates
# 1=large, 2=medium, 3=small internally
# R now treats rating as ordinal
```

Package Installation in R

To install a package in R, use the following command:

```
install.packages('chron') # "chron" is the name of the package
```

Once the package is installed, the command to load the installed package is library(). For example,

```
library(chron)
```

Reading and Writing Data in R

R supports various packages that allow developers to read data from various file formats and load them into objects or write data to various file formats. Here, we will discuss about how to read data from the CSV file and write data into the CSV file.

To read data from a CSV file and assign it to a dataframe "df," use the following command:

```
df <- read.csv("<file-name>", header=TRUE, sep=",")
```

Note here that

```
df : Name of the dataframe
<file-name>: Name of the source file with complete path to its
  location. If path is not specified, the file will be read
  from the working directory.
Header: Setting this parameter to TRUE indicates that the
  source file has the names of the columns to be read.
Sep: this parameter is used to indicate the delimiter in the
  source file, traditionally the delimiter for a csv is a
  comma (,)
```

To write the contents from the dataframe to a CSV file, use the following command:

```
write.csv(df, file = "<file-name>",row.names=FALSE)
```

Note:

```
df: the dataframe that you want to write to the CSV file.
<file-name>: target file name with complete path to its
  location. If path is not specified, the file will be written
  to the working directory.
row.names: this parameter will indicate whether you want to
  write the row.names (the index number of each row) to the
  target file.
```

Some commonly used functions that might come in handy in R programming are:

- length(object) # number of elements or components
- str(object) # structure of an object
- class(object) # class or type of an object
- names(object) # names
- c(object,object,...) # Combine objects into a vector
- cbind(object, object, ...) # Combine objects as columns
- rbind(object, object, ...) # Combine objects as rows
- rm(object) # delete an object
- colnames(object) # retrieve the column names of a matrix like object (matrix, dataframe, etc.)

Now consider some general programming examples.

Using the FOR Loop in R

The for loop in R can be implemented with the traditional syntax

```
# Creating a vector and assigning values to it
new <- c(2010,2011,2012,2013,2014,2015)
# for loop to display the values in the vector
for (year in new){  print(paste("The year is", year))}
```

Using the WHILE Loop in R

We will be using the same vector "new" that was created for the previous example.

```
# creating an increment counter
i <- 2010
# while loop to display the values in the vector
while (i < 2016) {print(i); i = i+1}
```

Using the IF-ELSE Statement in R

We will try to determine if the years mentioned in the vector "new" is a leap-year or not. For this we first need to define a function which checks the same.

```
# Function definition
is.leapyear=function(year){
   return(((year %% 4 == 0) & (year %% 100 != 0)) | (year %%
      400 == 0))
}
# Using the for loop and the if-else statements to determine
  the result
# the paste0 function is used to print different types of data
  together, both text and variable value
```

```
for (year in new){
  if (is.leapyear(year)==TRUE){
    print(paste0("The year ",year," is a leap year"))
  }
  else {
    print(paste0("The year ",year," is not a leap year"))
  }
}
```

PROGRAMMING WITH PYTHON

Now let's look at some basics of the Python programming. Readers are recommended to execute the provided scripts here in the Python terminal or Jupyter Notebook.

First, we'll display some text on the screen using the print command in Python.

```
print("Mary had a little lamb,")
print("its fleece was white as snow;")
```

Next, let's declare a variable in Python to hold an integer value and perform basic mathematic operation on the variable.

```
v = 2 # declare a variable in Python
print(v)
# Perform some arithmetic operations on the Python variable
v = v * 5
print(" The value of the variable v is::", v)
```

Python variables can also be used to store strings. For example,

```
word1 = "Good" # Assign the word Good to the Python variable
  word1
word2 = "morning"
word3 = "to you too!"
```

To combine the string variables and display them as sentence, in the screen type the following commands.

```
print(word1, word2)
```

Note here that when word1 and word 2 are printed on the screen there is a white space between word 1 and word 2. The output should be "Good morning."

```
sentence = word1 + " " + word2 + " " + word3
```

Note here that when you assign the value of the string variable to another string variable you need to add a white space. The string variables are concatenated using the + sign.

```
print(sentence)
```

Now let's see an example of WHILE loop implementation in Python

```
a = 0 # initialize the value of the loop variable
while a < 10: # loop till a reaches the value of 10
     a = a + 1 # Increment a to make sure that the loop
        terminates
     print(a) # print the value of the variable a in each
        iteration.
```

Note that the two lines below the WHILE loop should be typed at a tab distance (indentation) to indicate that those lines have to be executed within the WHILE loop

Next, consider another example of a WHILE loop in Python. Here, note that the final print command is outside the while loop and is executed when the while loop terminates.

```
v = 5
while v < 20:
      v= v + 5 #this line is part of while loop
      print(v) # this line is part of while loop
print("End of while loop")
```

Next, consider an example of a FOR loop in Python.

```
for cnt, value in enumerate([10,20,30]):
print (cnt,value)
```

Now consider an example of a FOR loop in combination with the if-else statements. Note the indentations used here to represent the different blocks of the loop.

```
for n in range(2, 10):
    for x in range(2, n):
        if n % x == 0:
            print( n, 'equals', x, '*', n/x)
            break
    else:
        # no factors were found hence control is transferred
           to else block
        print(n, 'is a prime number')
```

Consider an example of the use of the break statement. Again, note the use of the indentations here to represent the different blocks.

```
count = 0
while True:
     count += 1
     if count > 5:
     break
     print (count)
# This block prints integers from zero to 5
```

Now let's use an example to illustrate the use of the continue statement. This block prints only the odd numbers between 0 to 10.

```
count = 0
while count < 10:
     count += 1
     if count % 2 == 0:
          continue
     print (count)
```

Now let's try to write a simple function to accept user inputs and print statements on the screen based on the user inputs. If the user enters 1 then print *Tea* or print *Coffee* if the user enters 2.

```
choice = int (input ("Enter 1 for Tea and 2 for Coffee:: "))
```

Now consider an example to illustrate a simple try-except block with a single argument. The try clause is used to raise a user-defined exception, and the except block catches it and handles as per the user definition.

```
try:
     raise MyE('Sample error')
except MyE as e:
     print (e)
```

The *del* statement can be used to remove names from namespace or remove items from collection as demonstrated below.

```
d = {'aa': 111, 'bb': 222, 'cc': 333}
print (d)
```

The output should be {'aa': 111, 'cc': 333, 'bb': 222}. Now let's perform the following operation.

```
del d['bb']
print (d)
```

The output should be {'aa': 111, 'cc': 333}.

Here's an example to demonstrate how functions can be defined and called (referenced) in Python. First, let's define the familiar *hello*() function.

```
def hello(): # define the function using def
print("hello") # statement inside the function
```

The *hello*() function can be called as

```
hello() # call the function
```

Consider an example to illustrate the passing of parameters to the function that concatenates the words and prints it in the screen

```
def funny_function(first_word, second_word, third_word):
    print ("The word created is: " + first_word + second_word
       + third_word)
```

The example provided below is a function that returns the concatenated string to the main program.

```
def funny_function1(first_word, second_word, third_word):
    return first_word + second_word + third_word
```

Now get the user input for each word.

```
word1 = input(" Enter the first word")
word2 = input(" Enter the second word")
word3 = input(" Enter the third word")
```

Now let's call the function with the parameters. Here, the returned concatenated string is captured by the variable *final_string*.

```
funny_function(word1, word2, word3)
final_string = funny_function1(word1, word2, word3)
print(final_string)
```

Consider an example that demonstrates the creation of the class with a constructor and object initialization.

```
class A(object):
    def __init__(self, name):
        self.name = name
    def show(self):
        print ('name: "%s"' % self.name)

a = A('dave')
a.show()
```

The output in this case should be the name *dave*.

Consider an example to read and write to files in Python. The following example converts all the vowels in an input file to upper case and writes the converted lines to an output file.

```
import string
def show_file(infilename, outfilename):
tran_table = string.maketrans('aeiou', 'AEIOU')
with open(infilename, 'r') as infile, open(outfilename, 'w')
  as outfile:
for line in infile:
```

```
    line = line.rstrip()
    outfile.write('%s\n' % line.translate(tran_table))
```

Finally, let us consider an example of creating *DataFrame* in Python.

```
import pandas as pd
import os
new_dataframe = pd.DataFrame)
   {
      "id_col": [100,200,300,400]
      "string_col": ['this', 'is', 'a column of', 'string']
      "float_col": [0.01, 0.99, 49.9, 51.1]
      "binary_col": [True, False, False, True]
   }
)
```

To view the created dataframe simply type

```
new_dataframe
```

In this chapter, the objective is merely to provide an exposure to programming in R and Python so that the readers can become familiar with the syntax used in R and Python scripts. In the latter chapters, R and Python scripts will be used interchangeably to illustrate the concepts discussed.

DATA WRANGLING AND ANALYTICS IN R AND PYTHON

Several essential packages in R and Python that will be very helpful for performing *data wrangling* (structuring and cleaning data) and for performing analytics are listed below. In the latter chapters several of these listed packages will be used to demonstrate their potential for data wrangling and analytics.

First, the following packages in R are quite useful.

Dplyr—This package can be used for performing data wrangling and data analysis. This package includes various functions for manipulating dataframes.

ggplot2—This package is popular for visualization. ggplot2 facilitates declarative creation of graphics. Using this package one can create aesthetically pleasing and elegant plots and graphs.

Tidyr—This package is popular for tidying the data.

Shiny—This is an interactive web application that allows embedding visualizations like graphs, plots, and charts. The interfaces are directly written in R and this package provides a customizable slider widget with built-in support for animation.

Caret—This package is popular for modeling complex regression and classification problems. This package has an extension well known as *CaretEnsemble*, which is used for combining different models.

E1071—This package is widely used for implementing analytics technique including clustering, Fourier Transform, Naive Bayes, SVM, etc.

Plotly—This package extends on the JavaScript library mainly focused for building interactive quality graphs. The created graphs can then be embedded on web applications quite easily using this package.

Popular packages in Python for Data Wrangling and Analytics include

Pandas—It is a popular open-source package that provides high-performance, easy-to-use data structures and data analysis tools. Pandas is a perfect tool for data wrangling. It is designed for quick and easy data manipulation, reading, aggregation, and visualization.

NumPy—The NumPy is a general-purpose array-processing package. This package provides high-performance multidimensional array objects and tools to work with the arrays. NumPy is an efficient container of generic multidimensional data.

SciPy—This package builds on the NumPy array object and is part of the stack which includes tools like Matplotlib, Pandas, and SymPy with additional tools. The SciPy library contains modules for efficient mathematical routines as linear algebra, interpolation, optimization, integration, and statistics.

Matplotlib—This library supports data visualization. Matplotlib is the plotting library for Python that provides an object-oriented API for embedding plots into applications.

Scikit Learn—This is a robust machine learning library. It features machine learning algorithms including SVMs, random forests, *k*-means clustering, dimensionality reduction, etc. The Scikit Learn package focuses only on modeling data and not on data manipulation.

Statsmodels—This package provides easy computations for descriptive statistics and estimation and inference for statistical models.

Plotly—is a graph plotting library. The Plotly graph library has a wide range of graphs that can be plotted including basic charts: Line, Pie, Scatter, Bubble, Dot, Gantt, Sunburst, Treemap, etc.

STRUCTURING AND CLEANING DATA

The structuring and cleaning of the structured data involve several steps including checking for the normality distribution of the data, detecting and dealing with any data having a bimodal distribution, and resolving issues related to outliers, missing values, skewed data, and duplicate records. In this section, a brief overview is provided to detect and deal with the data irregularities mentioned above [1, 4].

First, we'll discuss how to deal with a bimodal distribution and check for the normality distribution of the variables. A set of data has a *bimodal distribution* if the data is distributed in two clusters. We use a simple example using R to visualize a variable with a bimodal distribution and show how to transform this variable into having a normal distribution [1, 4]. The R code follows.

```
# Create a variable with Bimodal distribution
x = rnorm(100, mean = 10, sd = 2)
y = rnorm(100, mean = 20, sd = 2)
bimodal = c(x,y) # bimodal is a variable with Bimodal
  distribution
hist(bimodal)
```

Figure 1.2 shows the histogram of the variable *bimodal*, generated by the code, which has a bimodal distribution.

Now let's try a transformation step to convert a bimodal distribution into a normal distribution or *homoscedasticity*[5]. Here, we take the absolute value of the original variable after subtracting the mean value of the variable at each data point.

```
transformed <- abs(bimodal - mean(bimodal)) # This is the
  transformation step
```

The variable *transformed* should have a unimodal or should be normally distributed which can be confirmed using the Shapiro–Wilk test for the normality.

```
shapiro.test(transformed)

      Shapiro-Wilk normality test
           data:  transformed
      W = 0.98824, p-value = 0.09777
```

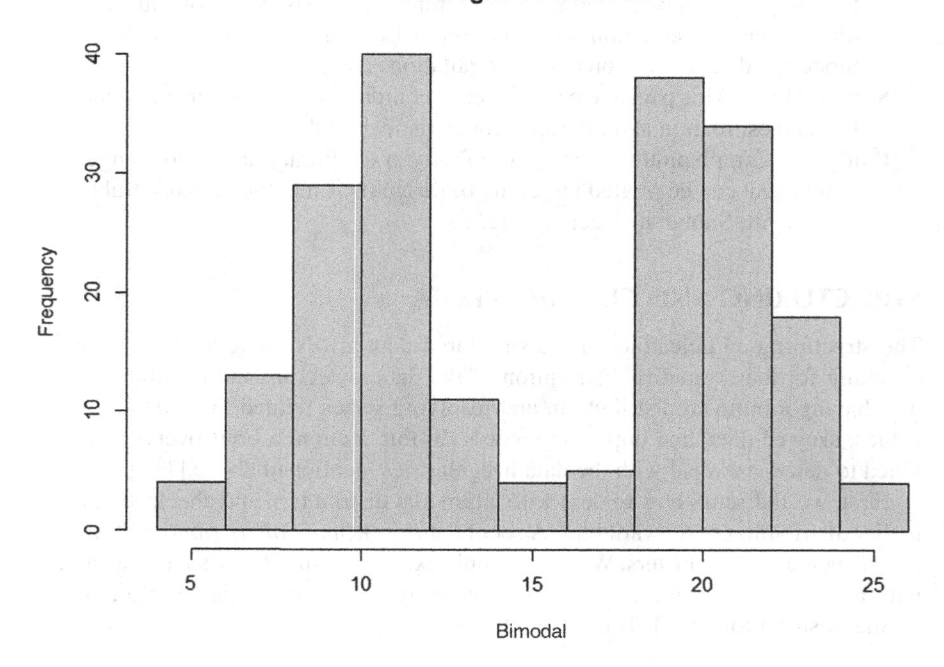

FIGURE 1.2 Variable *bimodal* with a bimodal distribution.

Note here that the Shapiro–Wilk normality test should result in a p-value > 0.05 after the transformation. Based on the p-value of 0.09777 we can infer that the *transformed* variable is normally distributed. This can also be confirmed by plotting the histogram of this variable (see Figure 1.3) [1, 4].

```
hist(transformed)
```

Now let's focus our discussion on a data transformation technique to address issues related to skewness. The skewness statistic can be used as a diagnostic. If the distribution of the variable is roughly symmetric, the skewness value would be close to zero. As the distribution becomes right skewed, the skewness statistic becomes larger. Similarly, as the distribution becomes more left skewed, the value becomes negative. In case the variable is right or left skewed then appropriate data transformation needs to be performed to remove the skewness [2].

There are several options for data transformation including replacing the data with the log, square root, inverse, etc. Once a suitable transformation is done, it is expected that the distribution of the predictor is entirely symmetric (normal distribution). However, evaluating the different options of transformation functions is cumbersome as each time a transformation is applied one needs to check for skewness. If the distribution is still skewed, then another transformation function needs to be applied [2].

In order to ease the steps, one can apply *Box–Cox Transformation*. Box and Cox proposed a family of transformations that is indexed by a parameter, denoted as λ [2]:

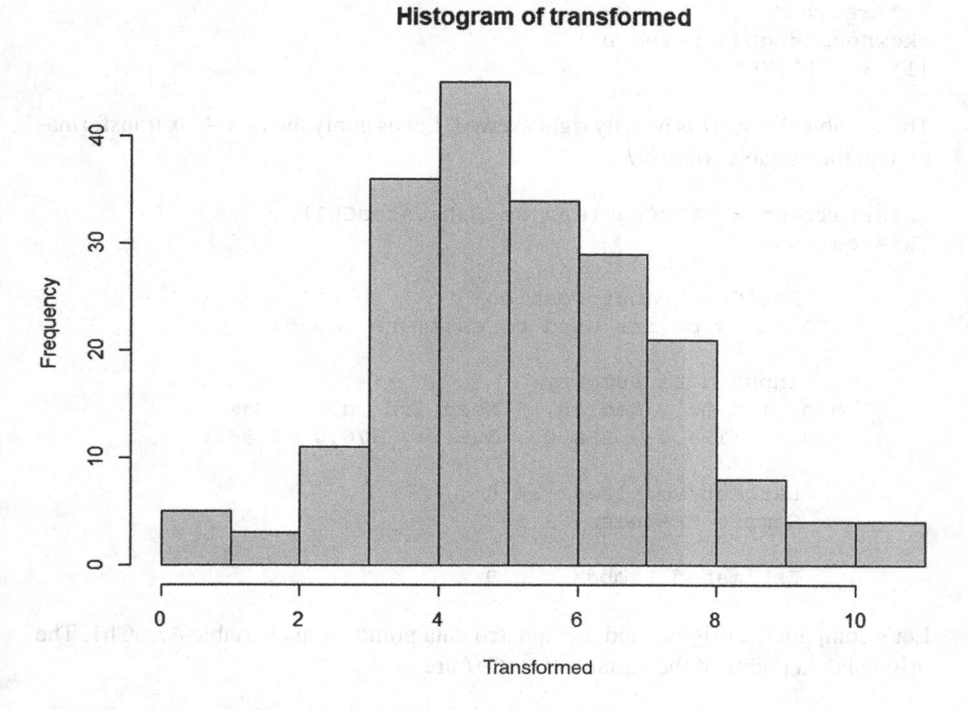

FIGURE 1.3 Variable *transformed* having a normal distribution.

$$x^* = \begin{cases} \dfrac{x^\lambda - 1}{\lambda}, & if\ \lambda \neq 0 \\ \log(x), & if\ \lambda = 0 \end{cases}$$

In addition to the log transformation, this family can identify square transformation ($\lambda = 2$), square root ($\lambda = 0.5$), inverse ($\lambda = -1$), and others in-between. Once the λ is estimated, the values (data points) of the variable x^* can be estimated using the above equation. Consider an example using the R to demonstrate the application of the Box–Cox transformation on a variable [2]

```
# install the "AppliedPredictiveModeling" package in R
Install.packages("AppliedPredictiveModeling")
# Now load the package AppliedPredictiveModeling
library(AppliedPredictiveModeling)
# In this package we will need access to the dataset
  segmentationOriginal
 data("segmentationOriginal")
# Partition the segmentationOriginal dataset to consider only
  those instances that have been labeled as "training".
segData <- subset(segmentationOriginal, Case == "Train")
# Install and load the caret package
Install.packages("caret")
library(caret)
# Now determine the skewness of the predictor variable
  "AreaCh1"
skewness(segData$AreaCh1)
[1] 3.52510745
```

The variable *AreaCh1* is heavily right skewed. Let us apply the Box–Cox transformation on the variable *AreaCh1*.

```
ChiAreaTrans <- BoxCoxTrans(segData$AreaCh1)
ChiAreaTrans

         Box-Cox Transformation
      1009 data points used to estimate Lambda

         Input data summary:
      Min. 1st Qu.  Median    Mean 3rd Qu.    Max.
      150.0   194.0   256.0   325.1   376.0  2186.0

         Largest/Smallest: 14.6
         Sample Skewness: 3.53

         Estimated Lambda: -0.9
```

Let's compare the original and the updated data points of the variable AreaCh1. The original data points of the variable *AreaCh1* are.

```
head(segData$AreaCh1)
[1] 819 431 298 256 258 358
```

The updated data points of the original variable *AreaCh1* are

```
predict(ChiAreaTrans, head(segData$AreaCh1))
[1] 1.108458 1.106383 1.104520 1.103554 1.103607 1.105523
```

The skewness value of the transformed variable is

```
> skewness(predict(ChiAreaTrans, head(segData$AreaCh1)))
[1] 0.4990754
```

Note that the distribution of the transformed variable looks more symmetric now indicating that the skewness of the variable has been treated to a greater extent [2].

Missing Data

What does it mean when data have missing values? Missing values usually appear as *NULL* values in a database or as empty cells in spreadsheet tables. Some flat-file formats use various symbols for missing values—for example, ARFF files use "?" symbol to represent missing values. These forms of missing values can be easily detected. However missing values can also appear as outliers or wrong data (i.e., out of boundaries) [6].

The primary task in dealing with missing values is to understand their characteristics. Missing values can be of different types namely [6].

- **Missing Completely at Random (MCAR)**: data are missing completely at random (MCAR) when the probability of a record having a missing value for an attribute does not depend on either the observed data or the missing data [6].
- **Missing at Random (MAR)**: data are considered missing at random (MAR), when the probability of a record having a missing value for an attribute could depend on the observed data, but not on the value of the missing data itself. Data which are incomplete only due to structural reasons are MAR [6].
- **Not Missing at Random (NMAR)**: missing data is considered as not missing at random (NMAR), when the probability of a record having a missing value for an attribute could depend on the value of the attribute. Missing data mechanism that is considered as NMAR is non-ignorable [6].
- **Structurally Missing***: these are data* that are missing for a logical reason. In other words, they are data that are missing because they should not exist [6].

Consider an example dataset to understand the characteristics of the Missing data shown in Table 1.1 [6].

TABLE 1.1
Table with Structurally Missing Data

ID	Children	Age of youngest child	Does the child go to school?
1	No		No
2	Yes	18	Yes
3	No		No
4	Yes	13	No
5	Yes	8	Yes

TABLE 1.2
Table with Missing Data for the *Income* Column

ID	Gender	Age	Income
1	Male	Under 30	Low
2	Female	Under 30	Low
3	Female	30 or more	High
4	Female	30 or more	
5	Female	30 or more	High

Here we can see that cell row 1, column 3 is missing data. This data is missing because the parents with ID = 1 do not have any children. So logically, the cell (row 1, column 3) should be empty or a missing value. This is a good example for structurally missing data [6].

Now consider Table 1.2. Here, the entry for Income in row 4 and column 4 is missing.

Let's ask ourselves "what is the likely income of the fourth individual?" The simplest approach to answering this question is to note that 50% of the other individuals have high incomes and 50% have low incomes. We could assume, therefore, that there is a 50% chance that the individual (she) has a high income and a 50% chance that she has a low income. This assumption is based on the fact that the missing value has an MCAR characteristic [6].

When we make this assumption, we are assuming that the missing data is completely unrelated to the other information such as *gender* and *age*. The MCAR assumption is rarely a good assumption. On the other hand, in the case of the MAR, we can assume that one can predict the value that is missing based on the other information (*gender* and *age*) provided in the table. A simple predictive model is that income can be predicted based on the variables *gender* and *age*. The missing value is for a female aged 30 or more, and the other females aged 30 or more have a *high income*. As a result, we can predict that the missing value should be *High*. Note that the idea of prediction does not mean we can perfectly predict a relationship. All that is required is a *probabilistic relationship* (i.e., we have a better-than-random probability of predicting the true value of the missing data) [6].

When data is missing at random, it means that we need to either use an advanced imputation method, such as *multiple imputation*, or an analysis method specifically designed for MAR data. Also, remember that the MAR is always a safer assumption than MCAR. This is because any analysis that is valid with the assumption that the data is MCAR will also be valid under the assumption that the data is MAR, but the opposite is not the case [6].

It may be the case that we cannot confidently make any conclusions about the likely value of missing data. For example, it is possible that people with very low incomes and very high incomes tend to refuse to answer. Or there could be some other reason we just do not know. This is known as NMAR and also as *nonignorable missing data*. It is common to include *structural missing data* as a special case of data that are NMAR. However, this misses an important distinction. *Structurally missing data* are easy to analyze, whereas other forms of NMAR data are highly problematic. When data is NMAR, it means that we cannot use any of the standard methods for dealing with missing data (e.g., imputation or algorithms specifically designed for missing values). If the missing data are missing not at random, any standard calculations would give the wrong answer [6].

Strategies for Dealing with Missing Data

Let's discuss some strategies for dealing with missing data. There are several mechanisms that are helpful in dealing with missing values, and each procedure has its own benefits and disadvantages. Popular strategies include [2, 6]:

Listwise deletion: simply omitting the cases (called *listwise deletion*) is the most frequently used method in handling missing data. Some researchers insist that it may introduce bias in the estimation of the parameters. However, if the assumption of MCAR is satisfied, a listwise deletion is known to produce unbiased estimates and conservative results. When the data do not fulfill the assumption of MCAR, listwise deletion may cause bias in the estimates of the parameters. If there is a large enough sample and the assumption of MCAR is satisfied, the listwise deletion may be a reasonable strategy [2, 6].

Regression imputation: this is the process of replacing the missing data with estimated values instead of deleting the cases with missing value. This approach preserves all cases by replacing the missing data with a probable value estimated by other available information. After all missing values have been replaced by this approach, the dataset is analyzed using the standard techniques for a complete data. This approach has several advantages because the imputation retains a great deal of data over the listwise or pairwise deletion and avoids significantly altering the standard deviation or the shape of the distribution [2, 6].

Replacing the missing value of the variable with the mean or median of that variable: This method replaces each missing value with mean of the attribute. The mean is calculated based on all known values of the attribute. Here, it is important to keep in mind that the mean is affected by the presence of outliers. So sometimes it seems natural to use the median instead, just to assure robustness. Also, remember that this method is usable only for replacing missing values for numeric attributes [2, 6].

Closest fit: In this technique, the missing value of an attribute, in an instance, is imputed by the value of the attribute belonging to the closest neighboring instance or instances. The closest neighbor of an instance is an instance that is at the closest proximity, which is determined by computing distances such as the *Manhattan* or the *Euclidian* distance [2, 6].

DATA DEDUPLICATION

With any dataset, but especially large ones derived from merged databases, there is the likelihood of duplicate data. It is usually very desirable to remove these duplications, that is to *dedup*[6] the data. Let's briefly discuss the data deduplication process [7].

In R programming, there are two functions namely *duplicated*() and *unique*(). The former function is used to identify duplicate records in the dataset and the latter is used for extracting the unique elements. Consider the following example R script [7]:

```
# First load the "tidyverse" package
library(tidyverse)
```

Now let us define a numeric vector containing elements namely 1,4,4,3,5,6,1

```
x <- c(1,4,4,3,5,6,1)
```

To find the position of the duplicate items in the vector use the *duplicated*() function

```
duplicated(x)
[1] FALSE FALSE  TRUE FALSE FALSE FALSE  TRUE
```

Duplicate items appear in the third and seventh location of the numeric vector *x*

To identify items that are duplicated, perform the following step

```
x[duplicated(x)]
[1] 4 1
```

To identify the unique items, perform the following steps

```
x[!duplicated(x)]
[1] 1 4 3 5 6

unique(x)
[1] 1 4 3 5 6
```

In many data analysis tasks, a large number of variables are being recorded or sampled. One of the important steps toward obtaining a coherent analysis is the detection of outlying observations. Outliers are generally considered an error or noise, but they carry important information. Detecting outliers are candidates for aberrant data that may otherwise adversely lead to model misspecification, biased parameter estimation, and incorrect results. Therefore, it is important to identify the outliers prior to performing any modeling and analysis. Here, we will discuss about two outlier detection techniques namely the Dixon's Q test and the Grubbs test [8, 9].

Dixon's Q test: The Dixon's Q test or simply the Q test is used for the identification and rejection of the outliers. This test should be used sparingly and never more than once in a dataset. To apply the Q test for bad data we need to first arrange the data in order of increasing values and then calculate Q as $Q = gap/range$ where *gap* is the absolute difference between the outlier in question and the closest number to it. If $Q_{calculated} > Q_{table}$ then we should reject the questionable point. The Dixon's Q test is more useful for small datasets with sample sizes less than 25 [8, 9].

Let's use a simple example to illustrate the Q test. Consider the following observed data points [8, 9]:

```
0.189, 0.167, 0.187, 0.183, 0.186, 0.182, 0.181, 0.184, 0.181,
  0.177
```

Consider the R script provided below

First, install the *outliers* package and load it

```
install.packages("outliers")
library(outliers)
```

Now create a numeric vector *data* containing the observed data points

```
data <- c(0.189, 0.167, 0.187, 0.183, 0.186, 0.182, 0.181,
  0.184, 0.181, 0.177)
```

Now call the *dixon.test* function in the *outliers* package

```
test <- dixon.test(data)
test

                    Dixon test for outliers
                         data:  data
                  Q = 0.5, p-value = 0.07653
      alternative hypothesis: lowest value 0.167 is an outlier
```

At $\alpha = 0.05$, we do not reject the null hypothesis (*p-value* = 0.0765 is greater than α) and conclude that the lowest value of the dataset, i.e., 0.167 is not an outlier [8, 9].

Grubb's test for outliers: this is a statistical test used for detecting outliers in a univariate dataset. This method is also known as maximum normed residual test. This test is based on the assumption of normality. That is, one should first verify that the data can be reasonably approximated by a normal distribution before applying the Grubb's test [8, 9].

The Grubb's test is an iterative process. This process detects one outlier at a time. The detected outlier is then expunged from the dataset and the test is iterated until no outliers are detected. However, multiple iterations change the probabilities of the detection, and the test should not be used for a small dataset, since it frequently tags most of the points as outliers [8, 9].

Grubb's test is defined for the hypothesis [8, 9]:

H_0: There are no outliers in the dataset.
H_a: There is at least one outlier in the dataset.

The Grubb's test statistic is defined as:

$$G = \frac{\max_{i=1,\ldots,N} |Y_i - Y'|}{s}$$

where Y' and s denote the sample mean and the standard deviation, respectively. This test is the largest absolute deviation from the sample mean in units of the sample standard deviation. This is the two-sided version of the test [8, 9].

The Grubb's test can also be defined as a one-sided test. To test whether the minimum value is an outlier, the test statistic is [8, 9]

$$G = \frac{Y' - Y_{\min}}{s}$$

where $Y_{\min}$ denotes the minimum value. On the other hand, to determine if the maximum value in the dataset is an outlier, the test statistic is [8, 9]

$$G = \frac{Y_{\max} - Y'}{s}$$

where $Y_{\max}$ denotes the maximum value.

For the two-sided test, the hypothesis of no outliers is rejected at significant level of if, α

$$G_{Grit} = \frac{N-1}{\sqrt{N}} \sqrt{\frac{t^2_{\frac{\alpha}{2N}, N-2}}{N-2+t^2_{\frac{\alpha}{2N}, N-2}}}$$

with $t^2_{\frac{\alpha}{2N}, N-2}$ denoting the upper critical value of the t-distribution with $N-2$ degree of freedom and a significant level of $\alpha/2N$ with α/N. Here, N is the sample size [8, 9].

Let's consider a simple example using the R script to illustrate the Grubb's test for outlier detection. Consider a reading of the heart rate of a person measured at six different intervals

```
80,     67,     66,     76,     78,     120
```

Out of the measured values it needs to be determined if the extreme value 120 is an outlier.

Now create a numeric vector *data* containing the observed data points

```
data <- c(80, 67, 66, 76, 78, 120)
```

Now call the *grubbs.test* function in the *outliers* package [8, 9]

```
test <- grubbs.test(data)
test

                    Grubb's test for one outlier
                           data: data1
           G = 1.95301, U = 0.08458, p-value = 0.008286
     alternative hypothesis: highest value 120 is an outlier
```

At $\alpha = 0.05$, we reject the null hypothesis (*p-value* = 0.0083 is less than α) and conclude that the extreme value 120 is an outlier.

SUMMARY

The focus of this chapter is to extensively discuss about the strategies for data collection, preprocessing, and cleaning. Data collection and cleaning is a prerequisite for predictive analytics. Therefore, the topics discussed here serve as a foundation for future chapters. Readers will have an opportunity to go through the basic programming concepts in R and Python. In addition to that, this chapter will also provide an inventory of all the packages that are available in R/Python for performing data wrangling, visualization, and analytics. Data cleaning has been discussed in detail highlighting a handful of strategies available for performing data transformation, dealing with missing values, taking care of data deduplication, and for detecting and removing outliers.

VIGNETTE Data Analytics and Developing Your Skill in Estimation

Nobel Prize winning Physicist Enrico Fermi (one of the fathers of atomic energy) was a proponent of informal estimation techniques to validate analytical results. Even though he was a world-class theoretical physicist and applied mathematician, he would use informal estimation to check his data analysis.

One famous example involves the testing of the atomic bomb. Fermi was present at the first detonation (at a safe distance). Shortly after the blast, he dropped some bits of paper in the air and observed how far they were blown by the blast wave. From this, he estimated the bomb to be 10K tons of TNT. After months of data analysis, it was found that the blast represented 18.6K tons. While Fermi was off by 86%, he was correct within an order of magnitude, providing a form of validation to the complex calculations.

Fermi used to challenge his classes with difficult problems for which there was ample data, but no direct approach to the solution—only estimation would do. For example, he would ask "how many piano tuners are there in Chicago?" Remember, this was long before there was the internet and the ability to search for such data. So the answer could only be obtained by library research.

His estimation approach was as follows. Collect the following data from almanacs and other library resources

1. Chicago has a population of about 3 million people (at that time).
2. Assume an average family has four members. Therefore, there are about 750,000 families in Chicago.
3. Assume one in five families owns a piano. Therefore, there are about 150,000 pianos in Chicago.
4. Suppose the average piano tuner could service four pianos every day of the week for five days (50 weeks per year). Then, in one year, a tuner could service 1,000 pianos.

So, $150{,}000/(4 \times 5 \times 50) = 150$, implying that there are about 150 piano tuners in Chicago.

This estimate is helpful to validate the results if we were to run a true data analysis, say by scrubbing web pages to get raw data on piano tuners in Chicago. For example, if such an analysis tells us that there are 10 tuners or a million, we know that the analysis is probably wrong, and we should revisit our approach [ref: https://www.grc.nasa.gov/www/k-12/Numbers/Math/Mathematical_Thinking/fermis_piano_tuner.htm].

Throughout this book, you will be working on rigorous mathematical formulae in data analytics. But sometimes there may be a mathematical truth behind some phenomenon of data but no logical explanation. Use your intuition and informal estimation techniques, where appropriate, to "check" the results to see if they make sense and challenge your assumptions if necessary.

EXERCISE

1. Discuss the challenges and opportunities associated with the Big Data analysis pipeline.

2. List the major tasks involved in the data preprocessing stage.

ID	Gender	Age	Income
1	Male	Under 30	Low
2	Female	Under 30	Low
3	Female	30 or more	Low
4	Female	30 or more	
5	Female	30 or more	High
6	Female	30 or more	High

3. In the above table, the missing cell (Row 4, Column 4) according to MCAR ("Missing Completely at Random") should have the entry
   A. Medium
   B. High
   C. Low
   D. Low and High
   E. Low or High

4. In the above table, the missing cell (Row 4, Column 4) according to MAR ("Missing at Random") should have the entry
   A. Medium
   B. High
   C. Low
   D. Low and High
   E. Low or High; it is inconclusive

5. Using the Grubb's test, determine if the value 146 of the univariate variable containing the following values (86, 92, 79, 64, 101, 121, 134, 94, 112, 36, 54, 146), is an outlier. Show all the steps of your determination

6. Consider the following values for a variable

   23, 45, 67, 78, 76, 65, 45, 32, 55, 66, 77, 110, 123, 456

   Is this variable skewed right or left? What is the skewness value? Apply the Box–Cox transformation on this dataset and write down the transformed values of this variable.

7. Which one of the 5 Vs of Big Data focuses on the validity and the correctness of the data?
   A. Velocity
   B. Variety
   C. Veracity
   D. Both A and B
   E. None

8. One of the following is not a phase of the Big Data analysis pipeline:
   A. Acquisition/Recording
   B. Ensuring data security
   C. Integration/Aggregation/Representation
   D. Analysis/Modeling
   E. Extraction/Cleaning/Annotation

9. Execute the following R script to create a bimodal variable (*bimodal*) and perform the Shapiro–Wilk normality test to confirm if the variable *bimodal* has a bimodal distribution

```
x = rnorm(60, mean = 10, sd = 5)
y = rnorm(60, mean = 60, sd = 5)
          bimodal =c(x,y)
```

10. What transformation steps will you employ to normalize the *bimodal* variable? After transformation, how will you confirm if the *bimodal* variable has a normal distribution?

NOTES

1 Data provenance means—where did the data come from? Do you trust it?
2 Global Positioning System
3 A data warehouse is a centralized repository of integrated data from one or more different sources.
4 With respect to data "dirty" refers to corrupt, inconsistent, or uncertain data.
5 Homoscedasticity means assuming that the underlying data of the transformed variable has equal or similar variance to the one being transformed.
6 Also sometimes spelled "dedupe."

REFERENCES

1. Agrawal, D., et. al. (2012). "Challenges and Opportunities with Big Data", *A Community White Paper Developed by Leading Researchers Across the United States*, retrieved from https://cra.org/ccc/wp-content/uploads/sites/2/2015/05/bigdatawhitepaper.pdf
2. Kuhn, M., Johnson, K. (2016). *Applied Predictive Modeling*. Springer.
3. Pratt, M. K. (2022). "How big data collection works: Process, Challenges, techniques", retrieved from https://www.techtarget.com/searchdatamanagement/feature/Big-data-collection-processes-challenges-and-best-practices, retrieved on March 9, 2022.
4. Rahm, E., Do, H.H., (2000). "Data Cleaning: Problems and Current Approaches", *IEEE Bulletin on Data Engineering*, 23(4), retrieved from http://dbs.uni-leipzing.de
5. "Data Preprocessing in Data Mining", retrieved from https://www.geeksforgeeks.org/data-preprocessing-in-data-mining/, retrieved on March 9, 2022.
6. Bock, T. (n.d.). "What are different types of missing data", retrieved from https://www.displayr.com/different-types-of-missing-data/, retrieved on December 16, 2019.
7. "Identify and Remove Duplicate Data in R", retrieved from https://www.datanovia.com/en/lessons/identify-and-remove-duplicate-data-in-r/, retrieved on March 9, 2022.
8. Barnett, V., Lewis, T. (1994). *Outliers in Statistical Data*. 3rd ed, Wiley.
9. Chawla, S., Sun, P., "Outlier Detection: Principles, Techniques and Applications", retrieved from http://www3.ntu.edu.sg/sce/pakdd2006/tutorial/chawla_tutorial_pakddslides.pdf, retrieved on November 1, 2015.

2 Mathematical Background for Predictive Analytics

In this chapter, we present the mathematical foundations required by data scientists to perform predictive analytics. The topics include basics concepts of linear algebra such as introduction to vectors; matrices, determinants, and equations for simple linear regression (SLR); dimensionality reduction techniques including Principal Component Analysis (PCA) and Singular Value Decomposition (SVD); and mathematical foundations for neural networks that will lay the foundations for the deep learning architectures discussed in the latter chapters.

BASICS OF LINEAR ALGEBRA

Linear algebra is a field of mathematics that is a prerequisite for understanding machine learning algorithms. Although linear algebra is a vast field with many complex theories, the fundamental notations and tools within the field are essential for machine learning practitioners. To fully understand how algorithms in predictive analytics work, it is important to master the foundations of linear algebra.

We will begin this chapter by introducing vectors, matrices, operations on matrices, determinants, and related concepts and then discuss one of the most basic, yet important concepts in data analysis, simple linear regression (SLR).

VECTORS AND MATRICES

A *vector* can be defined as a list of numbers (also known as *scalars*). For example, consider a vector $\vec{a} = [2,1]$. Graphically, you can think of this vector as an arrow in the *x*-*y* plane, pointing from the origin to the point at $x = 2$, $y = 1$, i.e., this vector extends 2 units in the *x*-axis and 1 unit in the *y*-axis. More generally, a vector $\vec{X} = [x_1, x_2, \dots, x_n]$ can be imagined as a point in an *n* dimensional space. The vector $\vec{a}$ noted above is just a point on an infinite-sized sheet of paper. Single component vector $\vec{b} = [6]$ is just the number 6 on the infinite number line. The vector $\vec{c} = [-1,2,7]$ is a point in 3-dimensional space. When the number of dimensions exceeds 3, however, it becomes difficult to actually visualize the vector, so it is usually not productive to try.

A vector has both magnitude and direction. You might recall from physics that the velocity vector of a moving object includes the speed (magnitude) it is moving at and

DOI: 10.1201/9781003278177-2

the direction in which the object is moving. In any case, the magnitude of the vector $\vec{X}$ is given as

$$\|\vec{X}\| = \sqrt{x_1^2 + x_2^2 + \ldots + x_n^2}.$$

The direction of a vector is denoted using a vector whose magnitude is 1 (known as a *unit vector*). To calculate the unit vector associated with a particular vector, we take the original vector and divide it by its magnitude. In mathematical terms, the unit vector associated with a vector $\vec{X}$ is denoted as $\hat{X} = \frac{\vec{X}}{\|\vec{X}\|}$. Again consider the vector $\vec{a} = [2,1]$. The magnitude of this vector is $\sqrt{2^2 + 1^2} = \sqrt{4+1} = \sqrt{5}$. Therefore, the unit vector associated with vector $\vec{X}$ is given as $\hat{X} = \left[\frac{2}{\sqrt{5}}, \frac{1}{\sqrt{5}}\right]$.

Vectors can be added and subtracted. In a graphical sense, we can think of adding two vectors together as placing two-line segments end-to-end, maintaining distance and direction. Let us consider an example of two vectors namely $\vec{a} = [2,1]$ and $\vec{b} = [3,2]$. The addition of these two vectors results in a third vector $\vec{c} = \vec{a} + \vec{b} = [2,1] + [3,2] = [2+3, 1+2] = [5,3]$. Similarly, in a vector subtraction $\vec{c} = \vec{a} - \vec{b} = [2,1] - [3,2] = [2-3, 1-2] = [-1,-1]$.

Now we wish to define the term *linearly independent*, which is a relationship between a set of vectors. A set of vectors is linearly independent if none of the vectors in the set can be created by any linear combination (or weighted sum) of the other vectors in the set.

For example, if two vectors point in different directions, even if they are not very different directions, then the two vectors are linearly independent. That is $\vec{c}$ is linearly independent of $\vec{a}$ and $\vec{b}$ if and only if it is impossible to find scalar values of α and β such that $\vec{c} = \alpha\vec{a} + \beta\vec{b}$.

The dot product of the two vectors $\vec{a}$ and $\vec{b}$ is given as $d = \vec{a} \cdot \vec{b}$. The dot product can also be geometrically represented as $d = \vec{a} \cdot \vec{b} = \|\vec{a}\| \cdot \|\vec{b}\| \cos\theta$ where the θ represents the angle between the two vectors. For example, the dot product of the two vectors $\vec{a} = [2,1]$ and $\vec{b} = [3,2]$ is $d = \vec{a} \cdot \vec{b} = [2,1] \cdot [3,2] = [2*3, 1*2] = [6,2]$.

Based on the understanding of the dot product we define the *orthogonality* of vectors. Two vectors are orthogonal to one another if the dot product of those two vectors is equal to zero. This happens when the angle between the vector is 90 degrees, i.e., $\theta = 90^o$ resulting in $\cos(90) = 0$.

A *matrix*, similar to vector, is a collection of numbers. The difference is that the matrix is a table of numbers rather than a list. Mathematically, we can define a matrix as an array of numbers made up of rows and columns. An $m*n$ matrix is characterized by the number of rows, *m*, and the number of columns, *n*. For example, a *2 × 2 matrix A and matrix B* can be denoted as

$$A = \begin{bmatrix} a_{11} & a_{12} \\ a_{21} & a_{22} \end{bmatrix}$$

$$B = \begin{bmatrix} b_{11} & b_{12} \\ b_{21} & b_{22} \end{bmatrix}$$

Matrices can be added, subtracted, and multiplied on an element-by-element basis, as we did for the vectors. Let us consider the addition and subtraction of two matrices A and B resulting in matrices C and D, respectively, as shown below

$$C = A + B = \begin{bmatrix} a_{11} & a_{12} \\ a_{21} & a_{22} \end{bmatrix} + \begin{bmatrix} b_{11} & b_{12} \\ b_{21} & b_{22} \end{bmatrix} = \begin{bmatrix} a_{11} + b_{11} & a_{12} + b_{12} \\ a_{21} + b_{21} & a_{22} + b_{22} \end{bmatrix}$$

$$D = A - B = \begin{bmatrix} a_{11} & a_{12} \\ a_{21} & a_{22} \end{bmatrix} - \begin{bmatrix} b_{11} & b_{12} \\ b_{21} & b_{22} \end{bmatrix} = \begin{bmatrix} a_{11} - b_{11} & a_{12} - b_{12} \\ a_{21} - b_{21} & a_{22} - b_{22} \end{bmatrix}$$

Matrix multiplication is more complicated since multiple elements in the first matrix interact with multiple elements in the second to generate each element in the product matrix. For example, let us consider the multiplication of two matrices A and B resulting in matrix E as shown below

$$E = A * B = \begin{bmatrix} a_{11} & a_{12} \\ a_{21} & a_{22} \end{bmatrix} * \begin{bmatrix} b_{11} & b_{12} \\ b_{21} & b_{22} \end{bmatrix} = \begin{bmatrix} a_{11}b_{11} + a_{12}b_{21} & a_{11}b_{12} + a_{12}b_{22} \\ a_{21}b_{11} + a_{22}b_{21} & a_{21}b_{12} + a_{22}b_{22} \end{bmatrix}$$

Solved Example

Consider the following two $2 * 2$ matrices

$$A = \begin{bmatrix} 2 & 3 \\ 4 & 5 \end{bmatrix}$$

$$B = \begin{bmatrix} 1 & 6 \\ 5 & 5 \end{bmatrix}$$

Let's perform addition, subtraction, and multiplication on the above matrices

$$C = A + B = \begin{bmatrix} 2 & 3 \\ 4 & 5 \end{bmatrix} + \begin{bmatrix} 1 & 6 \\ 5 & 5 \end{bmatrix} = \begin{bmatrix} 2+1 & 3+6 \\ 4+5 & 5+5 \end{bmatrix} = \begin{bmatrix} 3 & 9 \\ 9 & 10 \end{bmatrix}$$

$$D = A - B = \begin{bmatrix} 2 & 3 \\ 4 & 5 \end{bmatrix} - \begin{bmatrix} 1 & 6 \\ 5 & 5 \end{bmatrix} = \begin{bmatrix} 2-1 & 3-6 \\ 4-5 & 5-5 \end{bmatrix} = \begin{bmatrix} -1 & -3 \\ -1 & 0 \end{bmatrix}$$

$$E = A * B = \begin{bmatrix} 2 & 3 \\ 4 & 5 \end{bmatrix} * \begin{bmatrix} 1 & 6 \\ 5 & 5 \end{bmatrix} = \begin{bmatrix} 2*1+3*5 & 2*6+3*5 \\ 4*1+5*5 & 4*6+5*5 \end{bmatrix} = \begin{bmatrix} 17 & 27 \\ 29 & 49 \end{bmatrix}$$

If r is a scalar, then the scalar multiple of the matrix A is $r * A$, which is the matrix whose columns are r times the corresponding columns in A. For example,

$$5*A=\begin{bmatrix}2*5 & 3*5\\4*5 & 5*5\end{bmatrix}=\begin{bmatrix}10 & 15\\20 & 25\end{bmatrix}$$

Remember that a matrix with only a single column (or row) is a vector. For example, every column of the matrix A is a vector. Both the columns of the matrix A can be represented as vectors $v1$ and $v2$.

$$A=\begin{bmatrix}2 & 3\\4 & 5\end{bmatrix}\text{contains } v1=\begin{bmatrix}2\\4\end{bmatrix}\text{ and } v2=\begin{bmatrix}3\\5\end{bmatrix}$$

Now consider the multiplication of the matrix with a vector. If the matrix G is of size $m \times n$, and u is a vector of size n, then the product of G and u, denoted by Gu, is the linear combination of the columns of G using the corresponding entries in u as weights.

Let us assume G is a $2*2$ matrix and the vector u is of dimension $2*1$

$$G=\begin{bmatrix}1 & 2\\3 & 4\end{bmatrix}\text{ and } u=\begin{bmatrix}2\\4\end{bmatrix}\text{ then } Gu=\begin{bmatrix}1*2+2*4\\3*2+4*4\end{bmatrix}=\begin{bmatrix}10\\22\end{bmatrix}$$

Note that the product Gu is defined only if the number of columns of the matrix G equals the number of entries in the vector u.

Important Properties: If A, B, and C are all $m*n$ matrixes, u and v are vectors of size $n*1$ and r is a scalar, then:

$$A(u+v)=Au+Av$$
$$A(ru)=r(Au)$$
$$A(B+C)=AB+AC$$
$$(B+C)A=BA+CA$$
$$r(AB)=A(rB)=(rA)B$$
$$(AB)C=A(BC)$$

If A is an $n*n$ matrix and k is a positive integer, then A^k (A to the power k) is the product of k copies of A, i.e., $A^k=\underbrace{AA\ldots A}_{k\ times}$

Suppose we have a matrix F of size $m*n$, then the transpose of F (denoted by F^T) is a $n*n$ matrix whose columns are formed from the corresponding rows of F.

$$\text{If } \mathrm{F}=\begin{bmatrix}1 & 2 & 3\\4 & 5 & 6\end{bmatrix}\text{ then } F^T=\begin{bmatrix}1 & 4\\2 & 5\\3 & 6\end{bmatrix}$$

Important Properties: If A and B are both $m*n$ matrixes, and r is a scalar, then:

$$\left(A^T\right)^T = A$$
$$(A+B)^T = A^T + B^T$$
$$(rA)^T = rA^T$$
$$(AB)^T = B^T A^T$$

An $n*n$ matrix A also known as a *square matrix* is said to be *invertible* if there is an $n*n$ matrix C such that $CA = I$ and $AC = I$ where I is an $n*n$ identity matrix. An *identity* matrix is a square matrix containing 1's on the diagonal and 0's everywhere. If the matrix A is invertible then the matrix C can also be represented as A^{-1}, i.e., the matrix C is the *inverse* of matrix A.

Important Properties: If A and B are both $n*n$ invertible matrix, then:

$$\left(A^T\right)^{-1} = \left(A^{-1}\right)^T$$

$$\left(A^{-1}\right)^{-1} = A$$

$$(AB)^{-1} = B^{-1}A^{-1}$$

An *orthogonal* matrix is a square matrix whose columns and rows are orthogonal unit vectors. That is, an orthogonal matrix is an invertible matrix. For an $n*n$ matrix A, $AA^T = A^T A = I$. It can be implied here that for an orthogonal matrix A, $A^T = A^{-1}$.

Determinant

The determinant of a square matrix can be viewed as a function whose input is a square matrix and whose output is a number. For example, given a $2*2$ square matrix A, we can define the *determinant* of A as *det(A)* as:

$$\det\begin{vmatrix} a & b \\ c & d \end{vmatrix} = ad - bc \text{ where } A = \begin{bmatrix} a & b \\ c & d \end{bmatrix}$$

Now given a $3*3$ square matrix, the determinant can be computed as

$$det\begin{vmatrix} a & b & c \\ d & e & f \\ g & h & i \end{vmatrix} = a\,det\begin{vmatrix} e & f \\ h & i \end{vmatrix} - b\,det\begin{vmatrix} d & f \\ g & i \end{vmatrix} + c\,det\begin{vmatrix} d & e \\ g & h \end{vmatrix}$$
$$= a(ei - fh) - b(di - fg) + c(dh - eg)$$
$$= aei + bfg + cdh - afh - bdi - ceg$$

In a similar manner, the determinants for the higher orders of the square matrix can be determined.

SIMPLE LINEAR REGRESSION (SLR)

Linear regressions are performed to model the relationship between two variables by fitting a linear equation to the observed data. A linear equation can be expressed as $\boldsymbol{Y} = \beta_1 * \boldsymbol{X} + \beta_0$ where $\boldsymbol{Y}$ is the dependent variable, or the response variable and $\boldsymbol{X}$ is an independent variable also known as predictors. Both the variables, i.e., $\boldsymbol{Y}$ and $\boldsymbol{X}$ should be continuous. The slope of the line is β_1, and β_0 is the intercept (the value of $\boldsymbol{Y}$ when $\boldsymbol{X} = 0$). For example, if β_1 is 2 and β_0 is 1.5, the equation of the straight line is given by $\boldsymbol{Y} = 2 * \boldsymbol{X} + 1.5$.

A real-world application of SLR would be if a physician used a linear regression model to find a relationship between the weights of individuals and their heights. In any case it is very important to first determine whether or not there is a relationship between the variable of interest before fitting the linear model. Generally, a scatter-plot can be used to determine if there is a relationship between the variables of interest and the strength of their relationship. Alternatively, a numerical measure of association between two variables also known as correlation coefficient can also be used. This measure indicates the strength of association by a value that ranges between −1 and 1 where 1 indicates the strongest possible association [1, 2].

The most common method for fitting a regression line is the *method of least-squares*. This method calculates the best-fitting line for the observed data by minimizing the sum of the squares of the vertical deviations from each data point to the line (if a point lies on the fitted line exactly, then its vertical deviation is 0). However, in the real world, the data between the input variable ($\boldsymbol{X}$) and the response variable ($\boldsymbol{Y}$) never follows a straight line. Thus, we need to find a way to estimate the value of β_1 and β_0 using the given input data to fit a regression line. The approach that is used to estimate the value of β_1 and β_0 is called Ordinary Least Squares (OLS). The objective of the OLS is to obtain the minimum value for the Sum of Squared Errors (SSE) which is given by [1, 2]

$$SSE = \sum_{i=1}^{N} \left[\left(Y_i - \left(\beta_1 X_i + \beta_0 \right) \right) \right]^2$$

where N is the size of the learning dataset, Y_i is an actual value, $\hat{Y}_i = \left(\beta_1 X_i + \beta_0 \right)$ is the predicted value, and, $Y_i - \hat{Y}_i$ is a residual error.

When the SSE = 0, the straight regression line fits the data points perfectly implying that the correlation coefficient between X and Y is either +1 or −1. If SSE > 0, the line does not go through each data point implying that the correlation coefficient between X and Y ranges between +1 and −1 [1, 2].

Note here that the β_1 is the *covariance* of X and Y and β_0 is the *variance* of X. β_1 and β_0 can be determined as

$$\beta_1 = \frac{\sum_{i=1}^{N} \left(\left(X_i - \bar{X} \right) \left(Y_i - \bar{Y} \right) \right)}{\sum_{i=1}^{N} \left(X_i - \bar{X} \right)^2}$$

$$\beta_0 = \bar{Y} - \left(\beta_1 - \bar{X} \right)$$

Here, $\overline{Y}$ is the mean value of Y variable and $\overline{X}$ is the mean value of X variable.

Example

Table 2.1 depicts the height (inches) and weight (pounds) measurements for 5 individuals. Based on this dataset, determine if there is a linear relationship between the weight of the individuals and their height using a simple linear regression model.

Using the R programming language script shown in Figure 2.1, let's create a scatterplot to see if there is a relationship between the height and weight of all 5 individuals. Here, we will consider height as the input variable (X) and weight as the response variable (Y).

You should be able to visualize the scatterplot as shown in Figure 2.2. The black line is the regression line.

The correlation coefficient between the height and weight is 0.558 indicating a moderate level of relationship between the input and the response variable.

Now let us compute the regression line using the R code shown in Figure 2.3.

Executing the R code should result in $b_1 = 0.078$ and $b_0 = 132.47$. Therefore, the equation of the simple regression line can be expressed as $weight = 0.078 * height + 132.47$.

Now that you have a good understanding of the basics of linear algebra let's discuss the Principal Component Analysis technique for dimensionality reduction.

TABLE 2.1
Height and Weight Measurements for 5 Individuals

Index	Height (Inches)	Weight (Pounds)
1	65.78	112.99
2	71.52	136.49
3	69.40	153.0
4	68.22	142.34
5	67.79	144.30

```
height <- c(65.78, 71.52, 69.40, 68.22, 67.79)
weight <- c(112.99, 136.49, 153.03, 142.34, 144.30)

# Let's plot the scatterplot
plot(height, weight, main = "Height-Weight relationship",
     xlab = "Height", ylab = "Weight",
     pch = 19, frame = FALSE)

# Add a regression line
abline(lm(weight ~ height), col = "blue")

# find the correlation coefficient
cor(height, weight)
```

FIGURE 2.1 R script for producing scatterplot of height–weight relationship.

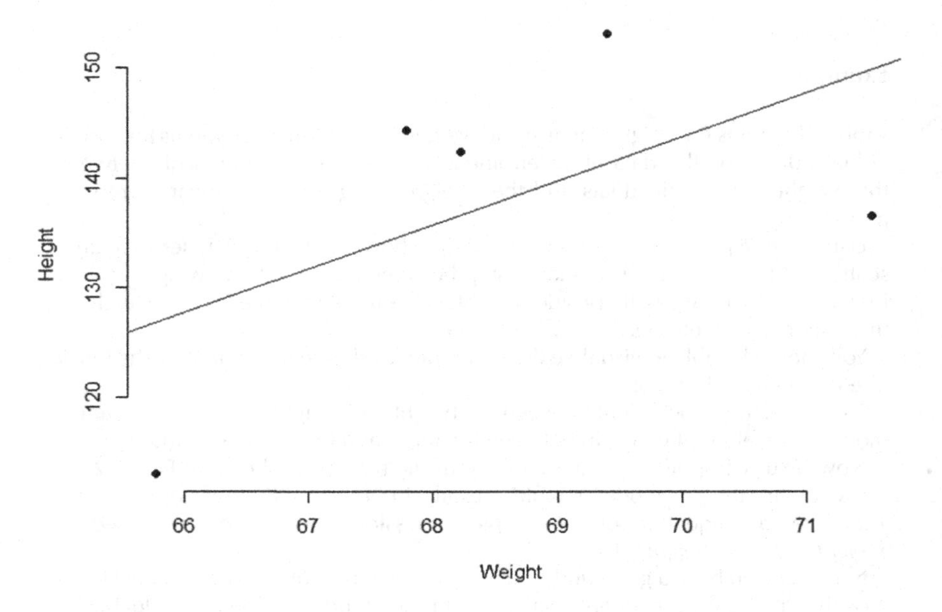

FIGURE 2.2 Scatterplot of height and weight relationship for 5 individuals.

```
height_mean = mean(height)
weight_mean = mean(weight)
height_var = sum((height - height_mean)**2)
height_var = sum((weight - weight_mean)**2)
covariance = sum((height-height_mean)*(weight-weight_mean))
b1 = covariance/height_var
b0 = weight_mean - b1*height_mean
```

FIGURE 2.3 R script to plot the regression line.

PRINCIPAL COMPONENT ANALYSIS (PCA)

Real-world datasets often have many variables, i.e., many dimensions. It is advantageous to reduce these to smaller sets of variables. For instance, in a consumer survey, there are many variables (questions) that are used to determine a small number of underlying concepts such as *customer satisfaction* with a service, *category leadership* for a brand, *luxury* for a product, etc. If we can reduce the data to its underlying dimensions, we can more clearly identify the relationships among concepts. Specifically, when there are data for many numeric variables, there is some redundancy among those variables. Redundancy means that some of the numeric variables are highly correlated with one another because they are measuring the same perspective. Therefore, it should be possible to reduce some of those original input numeric variables based upon a smaller number of principal components (PCs) that will account for most of the variance in the data of all the original input numeric variables. *Principal component analysis* (PCA) is an unsupervised technique that attempts to find uncorrelated linear dimensions that capture maximal variance in the data. *Unsupervised* means that there is no human interaction with the algorithm. PCA is a variable reduction procedure. The basic idea is to apply a linear transformation that defines a new space where the main axes are called the PCs [1, 2].

In short, PCA is used to [1, 2]:

1. Transform the original set of input numeric variables into a new set of variables, i.e., PCs, which explain the variance in the data of the original set of input numeric variables.
2. Each PC is a linear combination of all the original set of input numeric variables.
3. Total Number of PCs = Total Number of Original Input Numeric Variables.

If the original input numeric variables are $x_1, x_2, \ldots, x_n$ then $PC_i = w_{i1}x_1 + w_{i2}x_2 + \ldots + w_{in}x_n$, where w_{ij} is a component loading between PC_i and x_j for $1 \leq i, j \leq n$. Any input numeric variable with a relatively large magnitude of a component loading (w_{ii}) value (negative or positive) in any of the first few PCs is generally an input numeric variable that needs to be considered for modeling. The component loadings are analogous to correlation coefficients where squaring them gives the amount of explained variation. Therefore, the component loadings tell us how much of the variation in an input numeric variable is explained by this component[1, 2].

The principal components are extracted sequentially from the original set of input numeric variables as follows [1, 2].

- The 1st PC that explains the 1st most variance in the data.
- The 2nd PC that explains the 2nd most variance in the data and must be independent (i.e., zero correlation) of the 1st PC.
- The 3rd PC that explains the 3rd most variance in the data and must be independent of both 1st PC and 2nd PC.
- And so, on to additional PCs.

A *scree plot* usually displays the eigenvalues (Variances or Standard Deviation) associated with a principal component in descending order. Using the scree plots it is easier to visually assess which components explain most of the variability in the data, from which the PCs can be selected. An eigenvalue is the amount of variances explained by a PC. The highest eigenvalue indicates the highest variance in the data was observed in the direction of its principal component. From the scree plot, there are two ways that we can use to select the PCs, including *Eigenvalue-one Criterion* and *Proportion of Variance*[1, 2].

To summarize [1, 2]

- PCA tries to find uncorrelated linear dimensions that capture the maximal variance in the data.
- PCA recomputes a set of variables in terms of linear equations known as components that capture linear relationships in the data.

- The first component captures as much of the variance as possible from all variables as a single linear function.
- The second component captures as much variance as possible that remains after the first component.
- This procedure continues until there are as many components as there are variables.
- The objective is to retain a subset of components that can explain a large proportion of the variation in the data.

Let's explore PCA intuitively by looking at an example. We will use the programming language R to simulate data that is highly correlated. To do so we execute the following R code shown in Figure 2.4.

Here one can easily infer that xvar, yvar, and zvar are all correlated with each other. Let's visualize a bivariate plot and correlation matrix between xvar and yvar by adding the following code [1].

```
plot(yvar ~ xvar, data=jitter(my.vars))
cor(my.vars)
```

The result generates the scree plot shown in Figure 2.5.

From Figure 2.5, we can infer that xvar and yvar are both correlated (as indicated by the ellipse approximating a line). You can see that many points in this ellipse are clustered about this diagonal approximation line.

Now consider the correlation matrix shown in Table 2.2.

From the correlation matrix we can infer that xvar is highly correlated with yvar and less with zvar and also yvar is highly correlated with zvar. Now let's perform PCA and obtain the principal components using the following R code.

```
my.pca <- prcomp(my.vars)
summary(my.pca)
```

```
# Set the seed to create reproducible experiments
set.seed(10000)
# Create a sample of 50 values for xvar. Each value of xvar will range between 1 and 10.
xvar <- sample(1:10, 50, replace=TRUE)
# Copy the 50 values of xvar in to yvar
yvar <- xvar
# Let us replace 25 values of yvar
yvar[sample(1:length(yvar), 25)] <- sample(1:10, 25, replace=TRUE)
# Copy the 50 values of yvar in to zvar
zvar <- yvar
# Let us replace 25 values of zvar
zvar[sample(1:length(zvar), 25)] <- sample(1:10, 25, replace=TRUE)
# Bind xvar, yvar and zvar together
my.vars <- cbind(xvar, yvar, zvar)
```

FIGURE 2.4 R code to simulate highly correlated data.

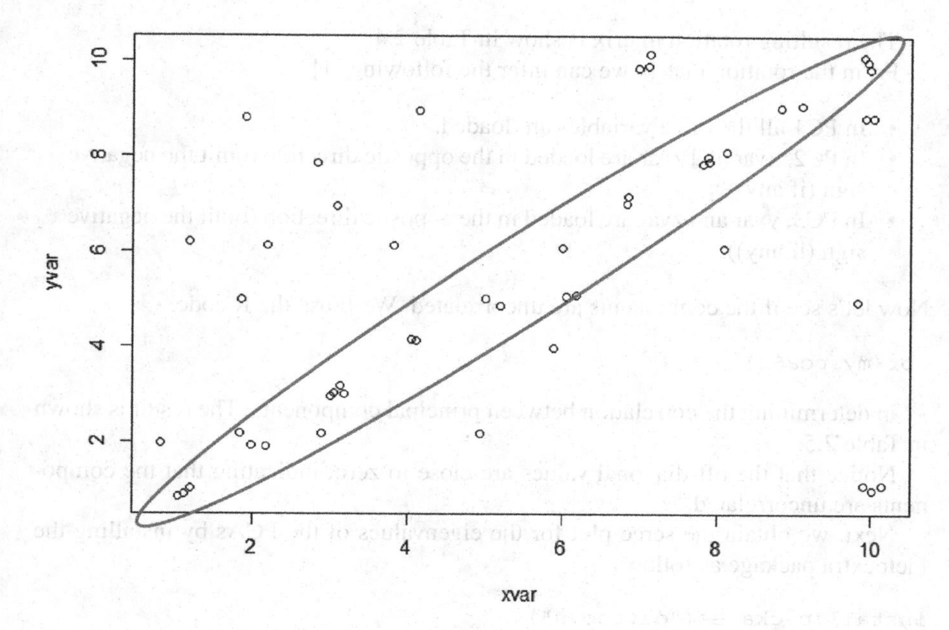

FIGURE 2.5 A bivariate plot between xvar and yvar.

TABLE 2.2
Correlation Matrix for xvar, yvar, and zvar

	xvar	**yvar**	**zvar**
xvar	1.000000	0.4825020	0.2191340
yvar	0.482502	1.0000000	0.6070264
zvar	0.219134	0.6070264	1.0000000

The results, shown in Table 2.3, show three PCs, their standard deviation, proportion of variance, and cumulative proportion.

The results indicate that PC1 explains 63.3% variance, and PC2 together with PC1 explains 89.5% (63.3% + 26.2%) variance in the dataset. Now let's print the rotation matrix using the following R code:

```
my.pca
```

TABLE 2.3
PCA Results for Dataset

Importance of components:

	PC1	PC2	PC3
Standard deviation	4.121	2.6503	1.6798
Proportion of variance	0.633	0.2618	0.1052
Cumulative Proportion	0.633	0.8948	1.0000

The resulting rotation matrix is show in Table 2.4.

From the rotation matrix we can infer the following [1]

- In PC1 all the three variables are loaded.
- In PC2, xvar and zvar are loaded in the opposite direction (omit the negative sign (if any)).
- In PC3, yvar and zvar are loaded in the opposite direction (omit the negative sign (if any)).

Now let's see if the components are uncorrelated. We'll use the R code:

```
cor(my.pca$x)
```

in determining the correlation between principal components. The result is shown in Table 2.5.

Notice that the off-diagonal values are close to zero, indicating that the components are uncorrelated.

Next, we obtain the scree plot for the eigenvalues of the PCAs by installing the factoextra package as follows:

```
install.packages("devtools")
library("devtools")
install_github("kassambara/factoextra")
library("factoextra")
fviz_eig(my.pca, geom="line")
```

The resultant plot is shown in Figure 2.6.

In the scree plot we see that the slope of the line is not very steep when we move from 2 to 3 dimensions. Therefore, we can conclude that only PC1 and PC2 are enough to explain a majority (~ 89%) of the variance in the data [1].

TABLE 2.4
Rotation Matrix for Dataset

Rotation (n × k) = (3 × 3):

	PC1	PC2	PC3
xvar	0.5591099	−0.7740077	−0.2971670
yvar	0.6734679	0.2149367	0.7072788
zvar	0.4835671	0.5955790	−0.6414425

TABLE 2.5
Correlation Values for Principal Components

	PC1	PC2	PC3
PC1	1.000000e+00	−6.192641e-17	8.416176e-17
PC2	−6.192641e-17	1.000000e+00	−2.472020e-16
PC3	8.416176e-17	−2.472020e-16	1.000000e+00

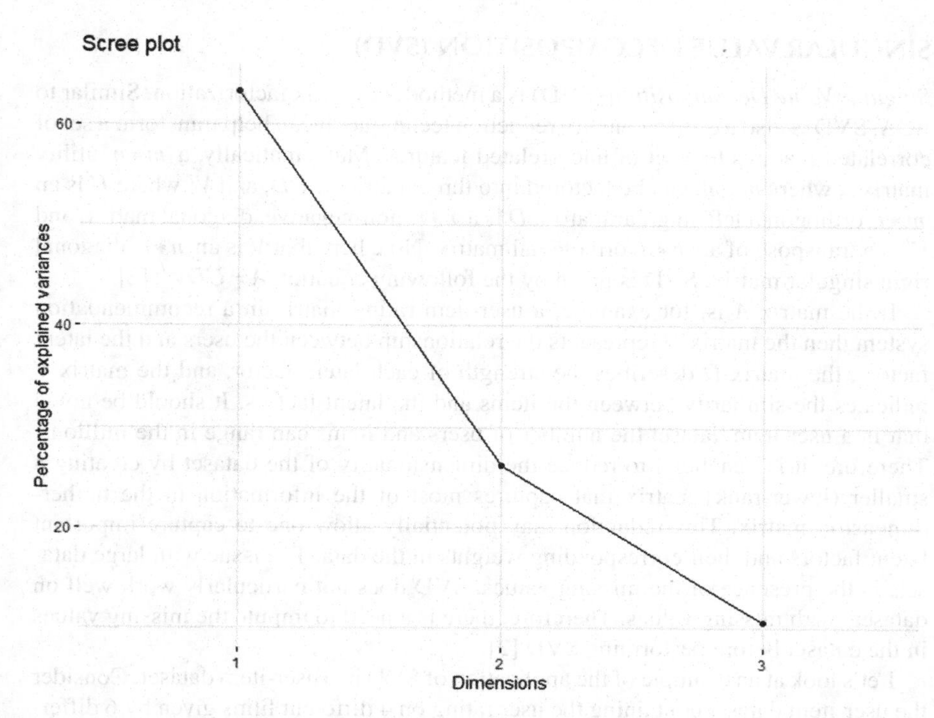

FIGURE 2.6 Scree plot of eigenvalues for PCAs.

VIGNETTE The Karhunen-Loève Transform

The Karhunen-Loève Transform (KLT) is a data transformation and analysis method often used for data compression and dimensionality reduction that is closely related to PCA. KLT is also referred to as Karhunen-Loève Decomposition (or Expansion), Principal (or Principle) Factor Analysis (PFA), SVD, Proper Orthogonal Decomposition (POD), Hotelling Transform, etc. KLT takes a given collection of data and creates an orthogonal basis for the data. KLT performs an orthogonal transformation $Y = \theta X$ on the vector X containing real values, where the resultant vector Y is free from any data correlation. The orthogonal matrix θ is composed of the eigenvectors of X satisfying the condition $\theta^T\theta = 1$ [3].

KLT has applications in almost any scientific field including image processing, data compression, studies of turbulence, thermal/chemical reactions, feed-forward and feedback control design applications, data analysis or compression (characterization of human faces, map generation by robots, and freight traffic prediction). Despite the favorable theoretical properties of the KLT, its usage for practical purposes is challenged by the fact that its basic functions are dependent on the covariance matrix. For image processing–related applications, the dependence of KLT on the covariance matrix requires that every image is recomputed and transmitted. In addition to that, the perfect decorrelation of KLT is not possible and there are no fast computation algorithms for its implementation [3].

SINGULAR VALUE DECOMPOSITION (SVD)

Singular Value Decomposition (SVD) is a method for matrix factorization. Similar to PCA, SVD is also a dimensionality reduction technique. It can help transform a set of correlated features to a set of uncorrelated features. Mathematically, a $m * n$ utility matrix A where $m > n$, can be factored into three matrices: UD, and V^T where U is an $m * r$ orthogonal left singular matrix, D is a $r * r$ non-negative, diagonal matrix, and V^T is a transpose of an $r * n$ orthogonal matrix. Note here that V is an $n * r$ diagonal right singular matrix. SVD is given by the following equation $A = UDV^T$ [2].

If the matrix A is, for example, a user-item rating matrix in a recommendation system then the matrix U represents the relationship between the users and the latent factors, the matrix D describes the strength of each latent factor, and the matrix V indicates the similarity between the items and the latent factors. It should be noted that in a user-item dataset the number of users and items can range in the millions. Therefore, it is beneficial to reduce the dimensionality of the dataset by creating a smaller (lower-rank) matrix that captures most of the information in the higher-dimension matrix. This reduction may potentially allow one to capture important latent factors and their corresponding weights in the data. The issue with large datasets is the presence of the missing values. SVD does not particularly work well on datasets with missing values. Therefore, there is a need to impute the missing values in the dataset before performing SVD [2].

Let's look at an example of the application of SVD in a user-item dataset. Consider the user-item dataset containing the user rating on 4 different films given by 6 different users shown in Table 2.6.

Applying SVD on the user-item dataset we can obtain the matrix U which shows the individual's (rows) loading on the 4 factors (columns) shown in Table 2.7.

TABLE 2.6
User Rating Data for 4 Films

	Film 1	Film 2	Film 3	Film 4
User 1	3	5	3	4
User 2	5	2	5	3
User 3	5	5	1	4
User 4	5	1	5	2
User 5	1	1	4	1
User 6	1	5	2	4

TABLE 2.7
Individual Rating Data for Four Factors

	[,1]	[,2]	[,3]	[,4]
[1,]	−0.4630576	0.2731330	0.2010738	−0.27437700
[2,]	−0.4678975	−0.3986762	−0.0789907	0.53908884
[3,]	−0.4697552	0.3760415	−0.6172940	−0.31895450
[4,]	−0.4075589	−0.5547074	−0.1547602	−0.04159102
[5,]	−0.2142482	−0.3017006	0.5619506	−0.57340176
[6,]	−0.3660235	0.4757362	0.4822227	0.44927622

The matrix D highlights the variance explained by each of the four factors (Table 2.8).

The total variance explained by the first two factors is given by

$$\frac{16.1204848 + 6.1300650}{16.1204848 + 6.1300650 + 3.3664409 + 0.4683445} = 85.29\%$$

and the matrix V in Table 2.9 shows the loading of each movie on a factor, i.e., the loading of film 4 on factor 1 is −0.1164526.

The R code in Figure 2.7 can be used to replicate these results [2].

You can tweak the code in Figure 2.7 to perform a similar analysis for other kinds of data.

TABLE 2.8
User Rating Variance (User 1) of the Four Factors

[1]	16.1204848	6.1300650	3.3664409	0.4683445

TABLE 2.9
The Loading Factors between Films 1, 2, 3, and 4

	[,1]	[,2]	[,3]	[,4]
[1,]	−0.5394070	−0.3088509	−0.77465479	−0.1164526
[2,]	−0.4994752	0.6477571	0.17205756	−0.5489367
[3,]	−0.4854227	−0.6242687	0.60283871	−0.1060138
[4,]	−0.4732118	0.3087241	0.08301592	0.8208949

```
# specify the user-item rating as
ratings <- c(3, 5, 5, 5, 1, 1, 5, 2, 5, 1, 1, 5, 3, 5, 1, 5, 4,2, 4, 3, 4, 2, 1, 4)
# Create a matrix of 6 rows and 4 columns
ratingMat <- matrix(ratings, nrow = 6)
# specify the row names which are users
rownames(ratingMat) <- c("User 1", "User 2", "User 3", "User 4", "User 5", "User 6")
# specify the column names which are the name of the films
colnames(ratingMat) <- c("Film 1", "Film 2", "Film 3", "Film 4")
# Display the matrix
ratingMat
# Perform SVD
svd <- svd(ratingMat)
# Display the matrix U, D and V
svd$u
svd$d
svd$v
# How much variance is explained by using just two factors
var1 <- sum(svd$d[1:2])
var2 <- sum(svd$d)
var1/var2
```

FIGURE 2.7 R code to replicate the results for the film rating example.

INTRODUCTION TO NEURAL NETWORKS

Neural networks are mathematical models that store information with the use of learning algorithms. The objective behind the use of the neural networks is to automatically learn and recognize complex patterns and make intelligent decisions based on the data. Neural networks are a popular framework to perform machine learning inspired by the brain biology of humans and other advanced organisms[2].

Neural networks model the relationship between a set of input signals and output signals using interconnected artificial *neurons* (or nodes) to solve complex pattern identification type problems [2].

Each neuron in the network has a set of inputs, each of which is associated with a specific weight (Figure 2.8).

The neuron computes a function on these weighted inputs. The neurons take a linear combination of weighted inputs and applies an activation function such as sigmoid, tanh, relu, etc. (which we will define shortly) on the aggregated sum. In this introductory section to neural networks, we will discuss about the single neuron which is also referred to as *perceptron* [2].

Very briefly, a perceptron is the most fundamental unit or a building block of a neural network. In a perceptron, the input signals are combined after multiplying them with different weights and are fed into the perceptron along with a bias element. Within the perceptron, the net sum is calculated as sum of weights and input signal and a bias element, then, the net sum is fed into a non-linear activation function [2].

The simplest neural network or perceptron shown in Figure 2.8 consists of n input signal. The process of passing the data through the perceptron is also known as *forward propagation.* Across a neuron each input signals x_i is multiplied by its respective weights w_i. These weights represent the strength of each signal that is given as an input to the neuron and decide how much influence the given input signal has on the neuron's output. The weighted input signals are then summed as [2]:

$$\sum(w_1x_1)+(w_2x_2)+(w_3x_3)+\ldots+(w_{n-1}x_{n-1})+(w_nx_n)$$

If we assume $x=\left[x_1,x_2,x_3,\ldots,x_{n-1},x_n\right]$ and $w=\left[w_1,w_2,w_3,\ldots,w_{n-1},w_n\right]$ both as row vectors, then the above expression can be expressed as a dot product of the row vectors given as [2]

$$w.x=\sum(w_1x_1)+(w_2x_2)+(w_3x_3)+\ldots+(w_{n-1}x_{n-1})+(w_nx_n)$$

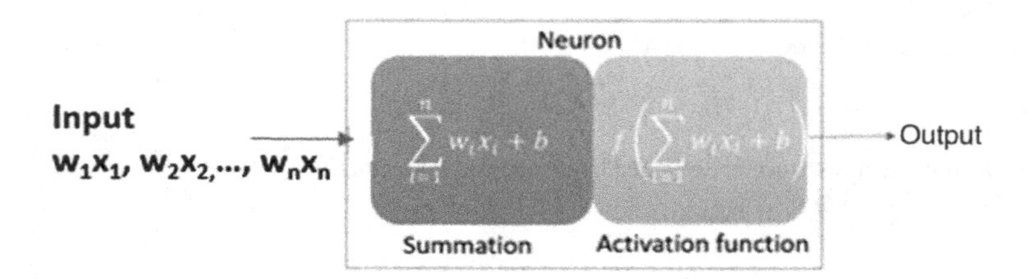

FIGURE 2.8 Model of an artificial neuron.

A bias b is then added to the above expression which is also referred to as an *offset* necessary to move the entire activation function to the left or right to generate the required output values [2].

$$w.x+b=\Sigma(w_1x_1)+(w_2x_2)+(w_3x_3)+\ldots+(w_{n-1}x_{n-1})+(w_nx_n)+b$$

After the application of the non-linear activation function the output of the perceptron can be expressed as

$$Y=f\left(\Sigma(w_1x_1)+(w_2x_2)+(w_3x_3)+\ldots+(w_{n-1}x_{n-1})+(w_nx_n)+b\right)$$

More specifically, the output of the perceptron is given as [2]

$$Y=f\left(\sum_{i=1}^{n}w_ix_i+b\right)$$

Now let's discuss the different types of non-linear activation functions used in neural networks.

Activation functions are the mechanisms by which a neuron processes information and passes it throughout the network. The activation function takes a single number and performs a certain fixed mathematical functional mapping on it. There are many different types of activation functions. The non-linear activation functions are more preferred for neural networks as the non-linearity aspect of the function makes it easy for the model to generalize or adapt with a variety of data and to differentiate between the outputs [2].

Sigmoid or logistic activation function takes a real number and converts it into a number in the range of 0 to 1. It is especially used for models where we have to predict the probability as an output. The sigmoid function curve looks like a S-shape curve and has a mathematical form

$$\sigma(x)=\frac{1}{1+e^{-x}}$$

The sigmoid function is differentiable and we can find the slope of the sigmoid curve at any two points. At the same time the sigmoid function is monotonic (non-increasing or non-decreasing) [2].

Tanh or hyperbolic tangent activation function is also like a sigmoid function with an s-shaped curve, but it converts a real-value number into the range of −1 to +1. The output is zero-centered, and its non-linearity is always preferred when compared to the non-linearity of the sigmoid function. Tanh function has a mathematical form given as

$$\tanh(x)=2\sigma(2x)-1$$

Tanh is a scaled sigmoid neuron. The tanh function is differentiable and is *monotonic* (continuously non-decreasing or non-decreasing) similar to the sigmoid function. Both the tanh and sigmoid activation function are popular with the feed forward neural networks [2].

Rectified Linear Unit (ReLU) is the most used activation function for neural networks and for deep learning. It computes the function $f(x) = \max(0, x)$. The ReLU is rectified from the bottom which means that if the input is less than zero then $f(x) = 0$. It converts a real-value number into the range of 0 to ∞. The ReLU function is monotonic. For all the negative values the $f(x)$ become zero immediately which decreases the ability of the model to fit or train from the data properly [2].

Softmax activation function is a function that converts a vector of K real values into a vector of K real values that sum to 1. The input values can be positive, negative, zero, or greater than one, but the softmax function transforms them into values between 0 and 1, so that they can be interpreted as probabilities. The softmax function is sometimes called as multi-class logistic regression. This is because the softmax is a generalization of logistic regression that can be used for multi-class classification, and its formula is very similar to the sigmoid activation function. Mathematically the softmax activation function is given by the expression

$$\sigma(\vec{z})_i = \frac{e^{z_i}}{\sum_{j=1}^{k} e^{z_i}}$$

where, $\vec{z}$ is the input vector to the softmax function made up of k elements $(z_0, z_1, \ldots, z_k)$, z_i is the element of the input vector, and k is the number of classes in the multi-class classifier [2].

SUMMARY

This chapter focuses on laying the mathematical foundations for performing predictive analytics. Various topics including the basic concepts of linear algebra such as introduction to vectors, matrices, determinants, and SLR are discussed here in detail. These mathematical foundations are very important to understand the theoretical concepts behind the dimensionality reduction techniques. The feature reduction techniques including the PCA and SVD which are vital preprocessing steps to perform predictive analytics were also discussed in detail. Finally, this chapter also lays the mathematical foundations for neural networks that are critical to the understanding of the deep learning architectures which will be discussed in the latter chapters.

EXERCISE

1. A company manufactures an electronic device to be used in a very wide temperature range. The company knows that increased temperature shortens the

lifetime of the device, and a study is therefore performed in which the lifetime is determined as a function of temperature. The following data are found:

Temperature in Celsius (t)	Lifetime in hours (y)
10	420
20	365
30	285
40	220
50	176
60	117
70	69

Perform a simple linear regression to determine the relationship between the change in the temperature and the lifetime of the device. Is there a positive correlation indicating that the increase in temperature shortens the lifetime of the device?

2. Based on the scatterplot what can you infer

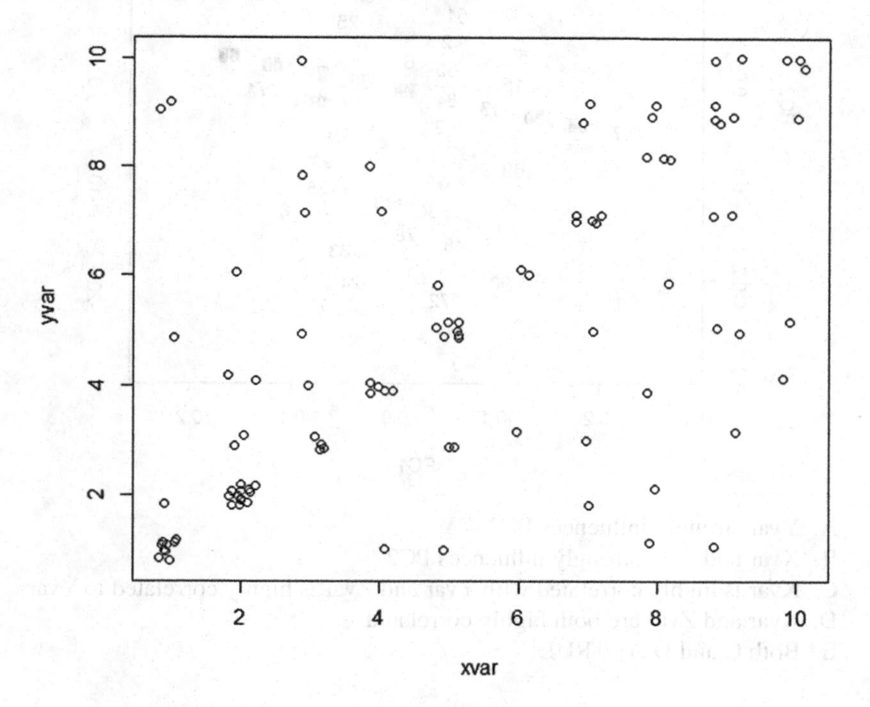

A. Xvar and Yvar are not at all correlated
B. Xvar and Yvar are correlated
C. Xvar and Yvar both have skewness
D. B and C
E. A, B, and C

3. Based on the table shown above, we can say that PC1 captures ______________ variance and PC2 together with PC1 captures ______________ variance in the dataset.
   A. 65.05, 89.23
   B. 65.05, 24.18
   C. 24.18, 65.05
   D. 10.77, 24.18
   E. 89.23, 100.00

4. Based on the biplot shown below one of the following statements is TRUE

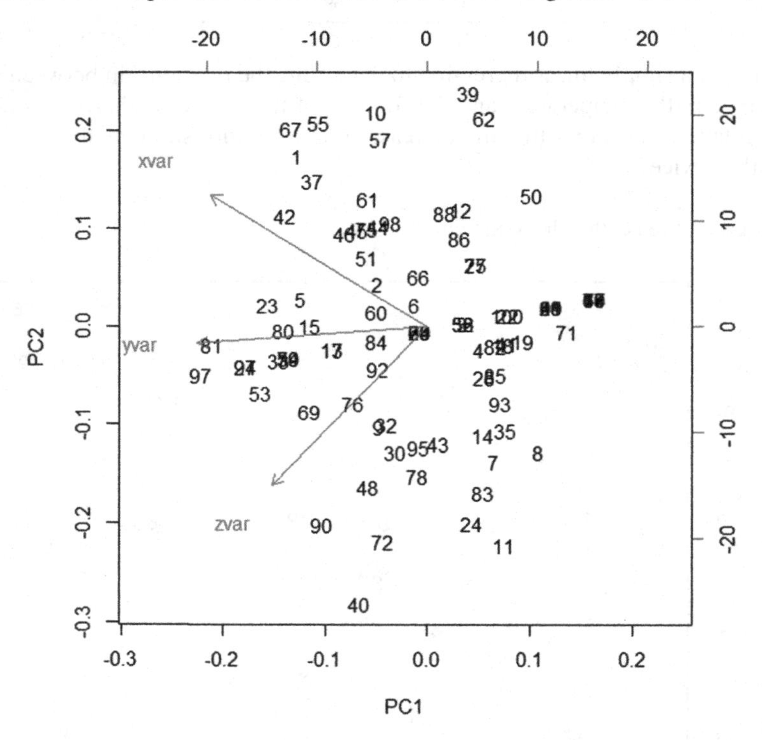

   A. Yvar strongly influences PC2
   B. Xvar and Yvar strongly influences PC2
   C. Xvar is highly correlated with Yvar and Zvar is highly correlated to Yvar
   D. Xvar and Zvar are both highly correlated
   E. Both C and D are TRUE

Based on the scree plot shown below answer the question 5

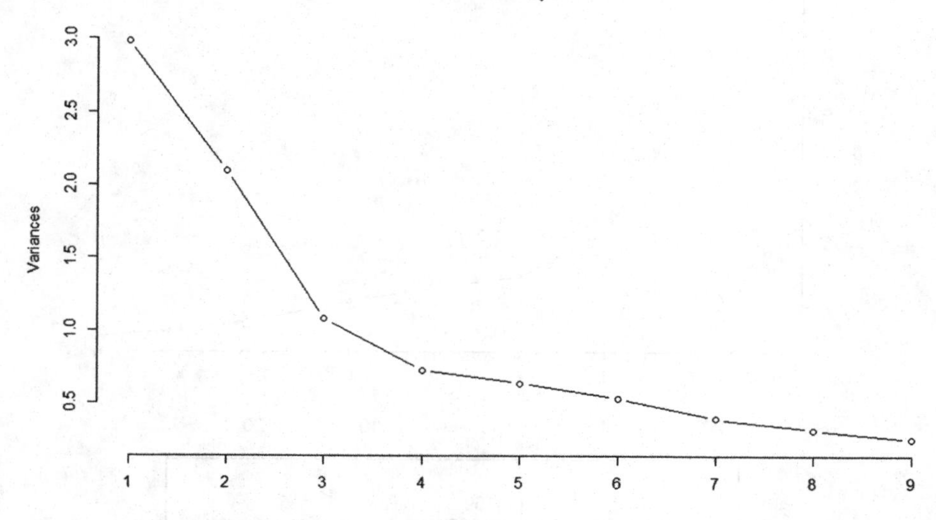

5. How many principal components (PC) should I choose?
   A. 1
   B. 9
   C. 7
   D. 3
   E. 4

Consider the Brand rating dataset for different brands (10 brands namely a, b, c, d, e, f, g, h, i, j) based on 9 perceptual adjectives namely perform, leader, latest, fun, serious, bargain, value, trendy, and rebuy. Thousand customers were provided with a survey instrument to rate each brand across 9 perceptual adjectives using a 10-point Likert scale where 1 is least and 10 is most. Once the survey result was obtained, PCA was performed on the responses of the survey.

```
> brand.pc <- prcomp(brand.sc[, 1:9])
> summary(brand.pc)
Importance of components:
                         PC1    PC2    PC3    PC4     PC5     PC6     PC7     PC8     PC9
Standard deviation     1.726 1.4479 1.0389 0.8528 0.79846 0.73133 0.62458 0.55861 0.49310
Proportion of Variance 0.331 0.2329 0.1199 0.0808 0.07084 0.05943 0.04334 0.03467 0.02702
Cumulative Proportion  0.331 0.5640 0.6839 0.7647 0.83554 0.89497 0.93831 0.97298 1.00000
```

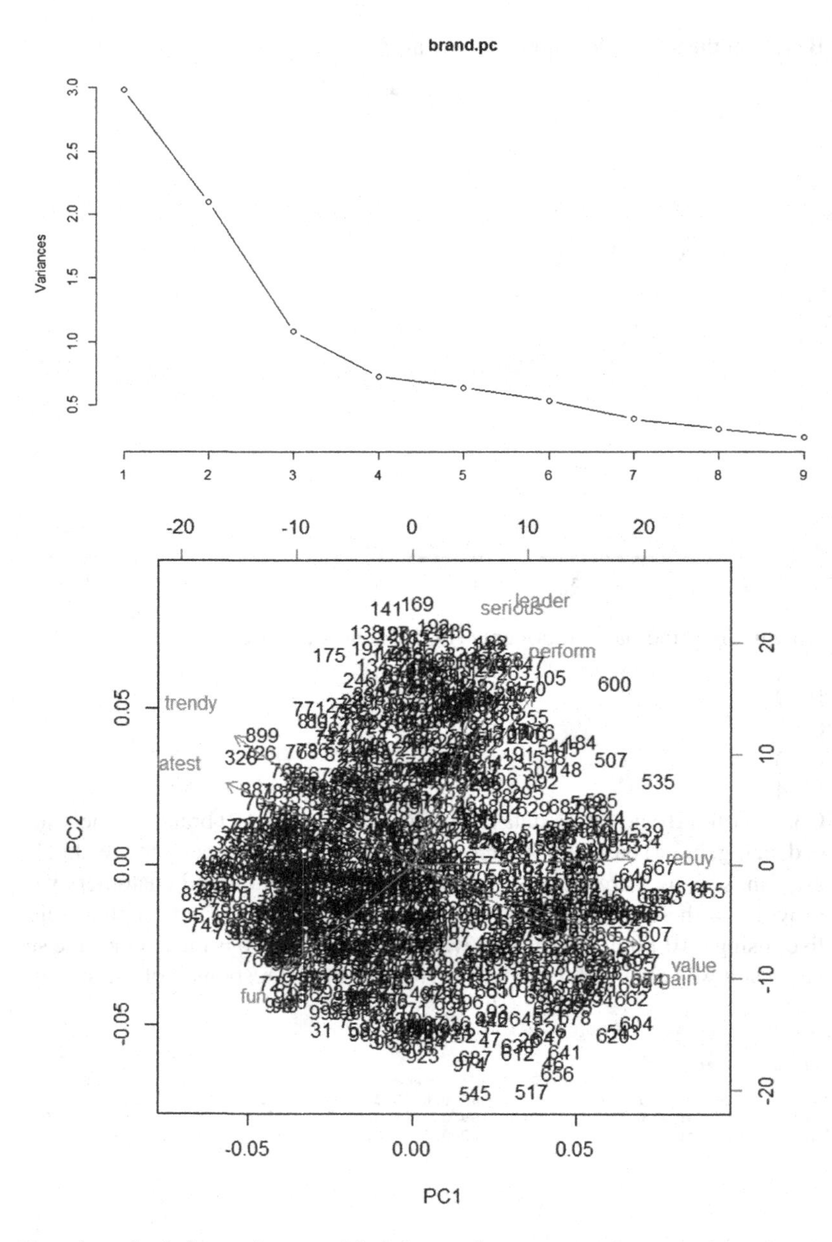

Based on the information provided above please respond to questions 6 and 7

6. The eigenvalues of PC1, PC2, and PC3 are
   A. 3, 4, 5
   B. 2.979, 2.096, 1.079
   C. 2.979, 4, 5
   D. 4, 2.096, 5
   E. 3, 4, 1.079

7. One of the following adjectives are positively correlated
   A. Perform, Fun
   B. Perform, Latest
   C. Perform, Serious, Leader
   D. Value, Bargain, Serious
   E. Value, Bargain, Perform, Serious

Consider the ratings of the user 1 to user 5 for 5 movies on a rating scale of 1 (worst)–5 (best) as shown below

	Movie 1	Movie 2	Movie 3	Movie 4	Movie 5
User 1	3	1	3	4	1
User 2	5	2	5	3	y
User 3	5	5	x	4	5
User 4	5	5	1	2	2
User 5	1	1	5	1	3

Upon performing SVD on the above user ratings I derived the following V matrix

	[,1]	[,2]	[,3]	[,4]	[,5]
[1,]	−0.5795873	−0.18456971	0.43911532	−0.3304055	0.5727325
[2,]	−0.4391825	−0.48864133	−0.22834662	−0.3488639	−0.6280925
[3,]	−0.3872198	0.85048790	−0.09874297	−0.2763658	−0.2015017
[4,]	−0.4124423	0.03147568	0.42637702	0.7445140	−0.3046345
[5,]	−0.3887794	−0.05332250	−0.75065839	0.3720848	0.3795683

Upon performing SVD on the above user ratings I derived the following D matrix

[1]	15.5472320	5.7871396	3.1360088	2.3311685	0.7236673

8. Based on the matrix above the Movie 4 is loaded on factor 1 by a magnitude of
   A. −0.38877
   B. −0.41244
   C. −0.33040
   D. 0.43911
   E. 0.57273

9. Using just the three factors the total variance explained is
   A. 0.55
   B. 0.44
   C. 0.775
   D. 1
   E. 0.889

10. SVD cannot be performed on the user rating matrix if it contains
    A. Lot of correlated variables
    B. Lot of outliers
    C. Lot of correlated variables and outliers
    D. Duplicated rows or columns
    E. Lot of missing values

REFERENCES

1. Chapman, C., Feit, E. (2015). *R for Marketing Research and Analytics*. Springer, ISBN 978-3-319-14436-8.
2. Dangeti, P. (2017). *Statistics for Machine Learning*. Packt Publishing Ltd., ISBN 978-1-78829-575-8.
3. Stathaki, T. (2014). "Digital Image Processing the Karhunen-Loeve Transform (KLT) in Image Processing", retrieved from https://www.commsp.ee.ic.ac.uk/~tania/teaching/DIP 2014/KLT.pdf, retrieved on February 16, 2022.

3 Introduction to Statistics, Probability, and Information Theory for Analytics

Statistics is an important and a necessary subject to gain a better understanding of the predictive analytics. However, it is so vast that it itself can be a complete book. Some of you may have studied probability and/or statistics in another course. Here, the attempt is to review and focus on the key concepts that are most necessary to understand the concepts of predictive analytics.

First, we introduce some basic terms [1].

Population—a complete list of all the observations in a study.

Sample—a subset of a population. It is usually a small portion of the population that is being analyzed.

Parameter—a measure that is calculated on the population.

Statistic—some measure that is calculated on a sample.

Mean—the arithmetic average of some collection of data, which is computed by taking the aggregated sum of all the data points divided by the number of data points. The mean is sensitive to outliers in the data.

Median—the midpoint of the data points. The mean is calculated by either arranging it in an ascending or descending order. If there are an odd number of data points, the median value is the number that is in the middle, with the same amount of data points below and above. If there is an even number of data points, the middle pair must be determined, added together, and divided by two to find the median value.

Mode—the most repeated data point in a set of data.

Measure of variation—describes the inconsistency or dispersion which is the spread of the values of a variable.

Range—the difference between the maximum and minimum value of the data points.

Variance—a measure of how far a set of data points is dispersed out from their mean or average value. For the set $X = (x_1, x_2, \ldots, x_N)$ containing N data

DOI: 10.1201/9781003278177-3

points with μ denoting the mean of the set X, the variance for the population and sample are given as

$$\text{population variance} = \frac{1}{N}\sum_{i=1}^{N}(x_i - \mu)^2 \text{ and sample variance} = \frac{1}{N-1}\sum_{i=1}^{N}(x_i - \mu)^2$$

Standard deviation—the square root of variance. It measures the spread of the statistical data.

$$\text{population standard deviation } (\sigma) = \sqrt{\frac{1}{N}\sum_{i=1}^{N}(x_i - \mu)^2}$$

$$\text{sample standard deviation (s)} = \sqrt{\frac{1}{N-1}\sum_{i=1}^{N}(x_i - \mu)^2}$$

Quantiles—identical fragments of the data. After arranging the data points in the ascending order, we can describe percentiles, deciles, quartiles, and so on.

Percentile—the percentage of data points below the value of the original whole data. The median is the 50th percentile.

Decile—the 10th percentile. The number of data points below the decile is 10% of the whole data.

Quartile—one-fourth of the data, and also is the 25th percentile.

Interquartile range—the difference between the third quartile and the first quartile. This measure is effective in identifying the outliers.

Let's consider a simple example to illustrate the basic concepts of statistics discussed above.

Consider a set S consisting of the following 10 data points

$$S = \{4,2,3,6,7,4,12,4,26,56\}$$

We can execute the R code shown in Figure 3.1 to provide examples of these statistical measures.

NORMAL DISTRIBUTION AND THE CENTRAL LIMIT THEOREM

The *normal distribution* is a probability distribution that is symmetric about the mean ($\mu = 0$, $\sigma = 1$), demonstrating that data near the mean are more frequent in occurrence than the data far from the mean [1].

The normal distribution formula is given as

$$f(x) = \frac{1}{\sigma\sqrt{2\pi}} e^{-\frac{1}{2}\left(\frac{x-\mu}{\sigma}\right)^2}$$

```
datapoint <- c(4,2,3,6,7,4,12,4,26,56)
data_mean = mean(datapoint); print(round(data_mean,2))
 Mean: 12.4
data_median = median (datapoint); print (data_median)
 Median: 5
func_mode <- function (inputdata) {
uniq <- unique(inputdata)
uniq[which.max(tabulate(match(inputdata,uniq)))]
}
datamode = func_mode (datapoint); print (datamode)
 Mode: 4
datavar = var(datapoint); print(round(datavar,2))
 Variance: 284.93
datastd = sd(datapoint); print(round(datastd,2))
 Standard deviation: 16.88
rangevalue<-function(x) return(diff(range(x)))
datarange = rangevalue(datapoint); print(datarange)
Range: 54
dataquantile = quantile(datapoint,probs = c(0.25,0.1,0.5));
print(dataquantile)
  Quartile  Decile  Median
    25%      10%     50%
    4.0      2.9     5.0
dataiqr = IQR(datapoint); print(dataiqr)
 Interquartile range: 6.75
```

FIGURE 3.1 R script and results illustrating various basic statistical measures.

The normal distribution curve appears as a bell curve. The normal distribution curve has a skewness of zero and the kurtosis[1] is 3.

The normal distribution is one of the most important concepts of predictive analytics and is the basis for many other important concepts. One of these follows.

The *central limit theorem* states that if you have a population with mean μ and standard deviation σ and take a sufficiently large random sample from the population with replacement, then the distribution of the sample means will be an approximate normal distribution [1].

Defining the normal distribution and introducing the central limit theorem allow us to continue our definitions and introduce other, slightly more sophisticated concepts of probability and statistics [1].

Hypothesis testing—is the process of making inferences about the population by conducting statistical tests on a sample. There are two types of hypotheses namely the null and the alternate hypotheses to validate whether an assumption is statistically significant or not.

***p*-value**—the value used when we have to decide either to accept or reject the null hypothesis. The p-value is the probability of obtaining a test statistic result is at least as extreme as the one that was actually observed. The p-value less than 0.05 is considered significant which means we have strong evidence to reject the null hypothesis.

Type I and II error—when performing inferences about the population from the samples there is possibility of making type I and II errors. Type I error occurs when we reject the null hypothesis when it is true and type II error occurs when we accept the null hypothesis when it is false. Larger sample size can reduce the probability of the occurrence of the type I and type II errors.

Chi-square—This is a test of independence between two categorical variables. Given two categorical variables X and Y, the chi-square test of independence determines whether or not there exists a statistical dependence between them. The test is usually performed by calculating $\chi^2$from the data. This statistic is a number that tells how much difference exists between the observed counts and the counts that would be expected if there were no relationship at all in the population [1].

$$\chi^2 = \Sigma \frac{(\text{Observed} - \text{Expected})^2}{\text{Expected}}$$

Let's use a simple example to illustrate the Chi-square test. In Table 3.1, all the species in the Iris dataset have been classified as either *small* or *big* in size based on whether the length of the petal is smaller or bigger than the median of all flowers.

The objective here is to determine if there is a relationship between the species and the size.

The null (H_0) and the alternate (H_a) hypothesis are:

H_0 = There is no relationship between the size and the species.
H_a = There is a relationship between the size and the species.

The R code for performing the Chi-square test is given in Figure 3.2.

The p-value = 2.2e-16 is lesser than 0.05. Hence, we have sufficient evidence to reject the null hypothesis and conclude that there is a relationship between the size and the type of species.

PEARSON CORRELATION COEFFICIENT AND COVARIANCE

The *Pearson Correlation Coefficient* or the *Pearson's product moment coefficient* is given as [1]

$$r_{A,B} = \frac{\sum_{i=1}^{n} (a_i - \bar{A})(b_i - \bar{B})}{(n-1)\sigma_A \sigma_B}$$

TABLE 3.1
Iris Flower Species Data

Size Species	Small	Large
Setosa	1	49
Versicolor	29	21
Virginica	47	3

```
# Create a vector d and then convert the vector into the matrix
d <- c(1, 29, 47, 49, 21, 3)
mat <- matrix(d, 3, 2)

# Now perform the Chi-Square test on the matrix
chisq.test(mat)

# The obtained output is

Pearson's Chi-squared test

data: mat
X-squared = 86.035, df = 2, p-value < 2.2e-16
```

FIGURE 3.2 R code to confirm hypothesis for Iris to size correlation.

where n is the number of tuples and $\bar{A}$ and $\bar{B}$ are the respective means of A and B, σ_A and σ_B are the respective standard deviations of A and B.

- If $r_{A,B} > 0$ then A and B are positively correlated. The higher the value, the stronger the correlation.
- If $r_{A,B} = 0$ then A and B are independent of each other.
- If $r_{A,B} < 0$ then A and B are negatively correlated.

Covariance is similar to correlation. The covariance is given as

$$cov(A,B) = \frac{\sum_{i=1}^{n}(a_i - \bar{A})(b_i - \bar{B})}{n} = E((A-\bar{A})(B-\bar{B})) = E(A.B) - \bar{A}\bar{B}$$

$$r_{A,B} = \frac{cov(A,B)}{\sigma_A \sigma_B}$$

where n is the number of tuples and $\bar{A}$ and $\bar{B}$ are the respective means or expected values of A and B, σ_A and σ_B are the respective standard deviation of A and B.

- Positive covariance: If $\text{cov}(A,B) > 0$, then if A is larger than its expected value, B is also likely to be larger than its expected value.
- Negative covariance: If $\text{cov}(A,B) < 0$, then if A is larger than its expected value, B is likely to be smaller than its expected value.
- Independence: $\text{cov}(A,B) = 0$ but the converse is not true:

Some pairs of random variables may have a covariance of 0 but are not independent. Only under some additional assumptions (e.g., the data follow multivariate normal distributions) does a covariance of 0 imply independence.

Let us consider an example to illustrate the concepts related to Pearson Correlation Coefficient and Covariance.

Suppose there are two stocks *A* and *B* having the following values in one week: (2, 5), (3, 8), (5, 10), (4, 11), (6, 14). If the stocks are affected by the same industry trends, will their prices rise or fall together?

$$E(A) = \frac{(2+3+5+4+6)}{5} = \frac{20}{5} = 4$$

$$E(B) = \frac{(5+8+10+11+14)}{5} = \frac{48}{5} = 9.6$$

$$cov(A,B) = \frac{(2*5+3*8+5*10+4*11+6*14)}{5} - (4*9.6) = 4$$

Thus, *A* and *B* rise together since $cov(A,B) > 0$.

BASIC PROBABILITY FOR PREDICTIVE ANALYTICS

Earlier we laid the foundations for the statistical concepts needed for performing analytics. Now, we'll look at the basic concepts in probability that are essential for performing analytics.

Probability is the branch of mathematics that describes how likely an event is to occur. The *probability* of any event is a real number between 0 and 1 (inclusive), where 0 means that the event would never occur and 1 means the event will certainly occur. Terms that are commonly used in connection to probability are *experiment*, *sample space*, and *outcome*. An experiment is a measurement process that produces quantifiable output. The outcome is a single result from an experiment and the sample space is the set of all possible outcomes from an experiment [1]. For example, consider throwing a die. Throwing a die (once) is an experiment. When a die is thrown, we could get any number that ranges from 1 to 6, i.e., the number 1, 2, 3, 4, 5, or 6. This is the sample space. Obtaining one number out of the six possible numbers is an outcome of the experiment. For example, when some die is thrown, we get the number 3.

The number of all possible outcomes may be finite, countably infinite, or constitute a continuum. Here, we will discuss only discrete, mainly finite, sample spaces. In discussing discrete sample spaces, it is useful to use basic set theory. Here, we recap some of the basic concepts in set theory using simple examples.

Let us assume the universal set *U* consists of nine elements given as $U = \{2,3,4,8,9,10,14,23,27\}$ and the sets *A* and *B* consists of three and four elements respectively given as $A = \{2,4,8\}$ and $B = \{3,4,8,27\}$ then we can define the following

$$\bar{A} = \{3,9,10,14,23,27\} \text{ and } \bar{B} = \{2,9,10,14,23\}$$

$$A \cup B = \{2,3,4,8,27\} \text{ and } A \cap B = \{4,8\}$$

$$A - B = \{2\} \text{ and } B - A = \{3,27\}$$

A *probability* is a number that reflects the chance or likelihood that a particular event will occur. Probabilities can be expressed as proportions that range from 0 to 1, and they can also be expressed as percentages ranging from 0% to 100%. A probability of 0 indicates that there is no chance that a particular event will occur, whereas a probability of 1 indicates that an event is certain to occur. The definition of probability states that if there are m outcomes in a sample space and are all equally likely of being the result of an experimental measurement, then the probability of observing an event (a subset) that contains S outcomes is given by S/m.

Consider the probability of drawing an ace from a standard deck of 52 playing cards. The sample space consists of 52 outcomes (i.e., each of the unique cards). The desired event (ace) is a set of 4 outcomes (ace of spades, clubs, hearts, and diamonds). Therefore, the probability of getting an ace is $4/52 = 0.0769$ or 7.69%.

Now let's discuss some axioms or rules of probability. Let S be a finite sample space, A an event in S. We define $P(A)$, the probability of A, to be the value of an additive set function $P()$ that satisfies the following three conditions [1]

- $0 \leq P(A) \leq 1$ for any event A in S (probabilities are real numbers on the interval $[0,1]$).
- $P(S) = 1$ (probability of some event occurring from S is unity).
- If A and B are mutually exclusive events in S, then $P(A \cup B) = P(A) + P(B)$ (The probability function is an additive set function).

CONDITIONAL PROBABILITY

The probability of an event is always defined with respect to the sample space S under consideration. Therefore, $P(A) = P\left(\frac{A}{S}\right)$, the conditional probability of event A relative to the sample space S [1, 2].

If A and B are any events in S and $P(B) \neq 0$, the conditional probability of A relative to B is

$$P\left(\frac{A}{B}\right) = \frac{P(A \cap B)}{P(B)}$$

In a similar way we can define $P\left(\frac{B}{A}\right) = \frac{P(A \cap B)}{P(A)}$

Let's consider a simple example to illustrate the concept of conditional probability. In a small company a proposal is made to change the color of the uniform from blue to brown. The proposal is submitted to the 20 uniform wearing workers for a vote. Upon polling the workers it is found that 12 are for and 8 are against. Suppose management wishes to discuss the vote with a small sample of two workers. What is the probability of randomly picking 2 workers that are against the proposal?

To answer this question, first we define:

Set A: all outcomes where the first worker is against
Set B: all outcomes where the second worker is against

Then $P(A)=\frac{8}{20}$ and $P\left(\frac{B}{A}\right)=\frac{7}{19}$ (Here we have to pick any 1 of the 7 left over workers who are against out of the remaining 19 workers)

Therefore,

$$P(A \cap B) = P(A)\,P\left(\frac{B}{A}\right) = \frac{8}{20} * \frac{7}{19} = \frac{14}{95}$$

So, the probability of picking two workers who are against the proposal is 14/19 ≈ 0.147

BAYES' THEOREM AND BAYESIAN CLASSIFIERS

Now let's discuss a very important theorem related to the concept of conditional probability, *Bayes' theorem* [2]. We'll introduce it by example.

Suppose an analyst for an auto dealer is asked to set up a model to predict the likelihood that certain individuals will buy a car based on age, income, and credit rating. That is, the auto dealer would like to be able to answer questions like the following:

Let X = 35-year-old customer, Jack, earning $75,000 per annum with a fair credit rating and H = Hypothesis that Jack will buy a computer today.

$$\text{Bayes theorem states that } P\left(\frac{H}{X}\right) = \frac{P\left(\frac{X}{H}\right)P(H)}{P(X)}$$

Here [2],

$P\left(\frac{H}{X}\right)$ = The conditional probability that Jack will buy a computer given the fact that the manager knows his age, income, and credit rating. This is also known as ***posterior probability*** of H.

$P(H)$ = Probability that Jack will buy a computer regardless of knowing his age, income, and credit rating. This is also known as ***prior probability*** of H.

$P\left(\frac{X}{H}\right)$ = Probability that Jack is 35 years old, earns $75,000 per annum, and has a fair credit rating given that he has already brought a computer from the store. This is also known as ***posterior probability*** of *X* or ***likelihood***.

$P(X)$ = Probability that Jack is 35 years old, earns $75,000 per annum, and has a fair credit rating. This is also known as ***prior probability*** of X or ***evidence***.

Therefore, in plain English we can say that $\text{posterior} = \frac{\text{prior} * \text{liklihood}}{\text{evidence}}$

So, in order to answer the posterior probability question for a specific individual, such as Jack, we would have to have data for the prior probability, the likelihood, and the evidence. Such a problem is one of Bayesian inference and Bayes theorem is the main tool in Bayesian inference [2].

Next, we will briefly discuss Bayesian classifiers. Bayesian classifiers are used to calculate the set of probabilities by counting the frequency and combination of values in a given dataset.

The probability model for the *Bayesian classifier* can be described as a conditional model as follows [2].

Assume a dataset D contains a set of tuples X where each tuple is an n dimensional attribute vector, i.e., $X:(x_1,x_2,x_3,\ldots,x_n)$ where x_i is the value of attribute A_i. Let there be m classes each denoted as $C_1,C_2,C_3,\ldots,C_m$. The Bayesian classifier predicts that the tuple X belongs to a class C_i if and only if $P\left(\frac{C_i}{X}\right) > P\left(\frac{C_j}{X}\right)$ for $i \le j \le m$ and $j \ne i$.

From the above, it is evident that the objective of the classifier is to maximize the product $P\left(\frac{X}{C_i}\right) * P(C_i)$ as $P(X)$ (denominator) is a constant.

The numerator part $P\left(\frac{X}{C_i}\right) * P(C_i)$ is actually $\left(\frac{x_1,x_2,x_3,\ldots,x_n}{C_i}\right) * P(C_i)$.

$$= P\left(\frac{x_1}{c_i}\right) * P\left(\frac{x_2,x_3,\ldots,x_n}{C_i,x_1}\right) * P(C_i)$$

$$= P\left(\frac{x_1}{c_i}\right) * P\left(\frac{x_2}{c_i,x_1}\right) * P\left(\frac{x_3,\ldots,x_n}{C_i,x_1,x_2}\right) * P(C_i)$$

$$= P\left(\frac{x_1}{c_i}\right) * P\left(\frac{x_2}{c_i,x_1}\right) * P\left(\frac{x_3}{c_i,x_1,x_2}\right) * P\left(\frac{x_4,\ldots,x_n}{C_i,x_1,x_2,x_3}\right) * P(C_i)$$

$$\ldots$$

$$= P\left(\frac{x_1}{c_i}\right) * P\left(\frac{x_2}{c_i,x_1}\right) * P\left(\frac{x_3}{c_i,x_1,x_2}\right) * \ldots * P\left(\frac{x_n}{C_i,x_1,x_2,x_3,\ldots,x_{n-1}}\right) * P(C_i)$$

This is where the "naïve" conditional independence assumptions come into play. Each feature x_i is independent of every other feature x_j for all $j \ne i$ which means that $P\left(\frac{x_i}{c_i,x_j}\right) = P\left(\frac{x_i}{c_i}\right)$

Therefore,

$$P\left(\frac{C_i}{x_1,x_2,x_3,\ldots,x_n}\right) = P\left(\frac{x_1}{c_i}\right) * P\left(\frac{x_2}{c_i}\right) * P\left(\frac{x_3}{c_i}\right) * \ldots * P\left(\frac{x_n}{c_i}\right) * P(C_i)$$

$$= P(C_i)\prod_{j=1}^{n} P\left(\frac{x_j}{c_i}\right) \text{ or, } \frac{1}{z}P(C_i)\prod_{j=1}^{n} P\left(\frac{x_j}{c_i}\right)$$

where Z (the evidence) is a scaling factor dependent only on $x_1,x_2,x_3,\ldots,x_n$, i.e., a constant if the values of the feature variables are known. Remember that this is because of the independence[2] assumptions [2].

The values of the attributes (A) can be either categorical or continuous. To compute $P\left(\frac{x_k}{C_i}\right)$ when the attribute values are categorical

$$P\left(\frac{x_k}{C_i}\right) = \frac{\text{the number of tuples of class } C_i \text{ in } D \text{ having the values } x_k \text{ for } A_k}{\text{the number of tuples of class } C_i \text{ in } D}$$

We will discuss the computation of the $P\left(\frac{x_k}{C_i}\right)$, where the attribute values are continuous, later.

Example

Consider the dataset D in Table 3.2 with the following tuples obtained from the customer database of a large consumer electronics retailer, such as Best Buy.

Based on this dataset we can compute the following probabilities:

There are two classes $C_1 = \textit{Buys computer} = \textit{Yes}$ and $C_2 = \textit{Buys computer} = \textit{No}$

The $P(C_1) = \frac{9}{14} = 0.642$ and $P(C_2) = \frac{5}{14} = 0.357$

Let's assume here that $age \leq 30$ is coded as youth in the dataset.

Therefore,

$$P\left(\frac{\text{age} \leq 30}{\text{Buys computer} = \text{YES}}\right)$$

$$= \frac{\text{number of tuples with buys computer} = \text{yes AND age} \leq 30}{\text{number of tuples with buys computer} = \text{yes}}$$

$$= \frac{2}{9} = 0.222$$

TABLE 3.2
Dataset for Customers of a Large Consumer Electronics Company

Rid	Age	Income	Student	Credit Rating	Buys Computer or Not?
1	Youth	High	No	Fair	No
2	Youth	High	No	Excellent	No
3	Middle-aged	High	No	Fair	Yes
4	Senior	Medium	No	Fair	Yes
5	Senior	Low	Yes	Fair	Yes
6	Senior	Low	Yes	Excellent	No
7	Middle-aged	Low	Yes	Excellent	Yes
8	Youth	Medium	No	Fair	No
9	Youth	Low	Yes	Fair	Yes
10	Senior	Medium	Yes	Fair	Yes
11	Youth	Medium	Yes	Excellent	Yes
12	Middle-aged	Medium	No	Excellent	Yes
13	Middle-aged	High	Yes	Fair	Yes
14	Senior	Medium	No	Excellent	No

In the similar way we can derive the following probabilities

$$P\left(\frac{\text{age} \le 30}{\text{Buys computer} = \text{NO}}\right) = \frac{3}{5} = 0.600$$

$$P\left(\frac{\text{Income} = \text{medium}}{\text{Buys computer} = \text{NO}}\right) = \frac{2}{5} = 0.400$$

$$P\left(\frac{\text{Income} = \text{medium}}{\text{Buys computer} = \text{YES}}\right) = \frac{4}{9} = 0.444$$

$$P\left(\frac{\text{Student} = \text{YES}}{\text{Buys computer} = \text{YES}}\right) = \frac{6}{9} = 0.667$$

$$P\left(\frac{\text{Student} = \text{YES}}{\text{Buys computer} = \text{NO}}\right) = \frac{1}{5} = 0.200$$

$$P\left(\frac{\text{Credit rating} = \text{fair}}{\text{Buys computer} = \text{YES}}\right) = \frac{6}{9} = 0.667$$

$$P\left(\frac{\text{Credit rating} = \text{fair}}{\text{Buys computer} = \text{NO}}\right) = \frac{2}{5} = 0.400$$

Now let's determine if the customer Jack who is 30 years old, a university student, earning $40,000 (medium salary range) and with a fair credit history will buy a computer or not. In order to determine this, we will have to compute the conditional probability

$$\begin{aligned} P\left(\frac{X}{\text{Buys computer} = \text{YES}}\right) &= P\left(\frac{\text{age} \le 30}{\text{Buys computer} = \text{YES}}\right) \\ &* P\left(\frac{\text{Student} = \text{YES}}{\text{Buys computer} = \text{YES}}\right) \\ &* P\left(\frac{\text{Income} = \text{medium}}{\text{Buys computer} = \text{YES}}\right) \\ &* P\left(\frac{\text{Credit rating} = \text{fair}}{\text{Buys computer} = \text{YES}}\right) \\ &= \frac{2}{9} * \frac{6}{9} * \frac{4}{9} * \frac{6}{9} = 0.044 \end{aligned}$$

Similarly, we can compute

$$P\left(\frac{\text{X}}{\text{Buys computer} = \text{NO}}\right) = P\left(\frac{\text{age} \leq 30}{\text{Buys computer} = \text{NO}}\right) * P\left(\frac{\text{Student} = \text{YES}}{\text{Buys computer} = \text{NO}}\right) * P\left(\frac{\text{Income} = \text{medium}}{\text{Buys computer} = \text{NO}}\right) * P\left(\frac{\text{Credit rating} = \text{fair}}{\text{Buys computer} = \text{NO}}\right) = \frac{3}{5} * \frac{1}{5} * \frac{2}{5} * \frac{2}{5} = 0.019$$

The objective is to find the class C_i that maximizes $P\left(\frac{X}{C_i}\right) * P(C_i)$

$$P\left(\frac{\text{X}}{\text{Buys computer} = \text{YES}}\right) * P(\text{Buys computer} = \text{YES}) = 0.044 * 0.642 = 0.028$$

$$P\left(\frac{\text{X}}{\text{Buys computer} = \text{NO}}\right) * P(\text{Buys computer} = \text{NO}) = 0.019 * 0.357 = 0.007$$

Since $P\left(\frac{\text{Buys computer} = \text{YES}}{\text{X}}\right) > P\left(\frac{\text{Buys computer} = \text{NO}}{\text{X}}\right)$ we can determine that John is a potential customer who will buy a computer.

As indicated before it is possible for the values of the attribute to be continuous. For example, consider the salary of the employees in an organization. This attribute takes numeric values which are continuous in nature. There are two ways to deal with the continuous values of the attributes. The attribute can be discretized. For example, the salary of the employee can be discretized as low (salary ≤ $30,000), medium ($31,000 to $75,000), and high (salary ≥ $100,000). The alternative way to deal with continuous data is to assume that the continuous values associated with each class are distributed according to a Gaussian distribution. Usually, a ***normal distribution*** is assumed, i.e., to estimate $P\left(\frac{X_i = x_k}{C_j}\right)$ for a value of the attribute X_i and for each class C_j [2]

$$P\left(\frac{X_i = x_k}{C_j}\right) = g\left(x_k; \mu_{ij}, \sigma_{ij}\right), \text{where } g\left(x; \mu, \sigma\right) = \frac{1}{\sigma\sqrt{2\pi}} e^{-\frac{(x-\mu)^2}{2\sigma^2}}$$

The mean μ_{ij} and the standard deviation σ_{ij} are estimated from the given dataset.

TABLE 3.3
Simple Dataset

Class	X_1	X_2
+	A	1.0
+	B	1.2
+	A	3.0
–	B	4.4
–	B	4.5

When there is a small number of samples in the training set or if the precise distribution of the data in the dataset is unknown then it is better to use the distribution method. Alternatively, when there is a large amount of training data the discretization method is the best choice.

Consider a simple dataset D shown in Table 3.3 to illustrate these concepts.
From this dataset, we can derive the following conditional probabilities:

1. Estimate $P(C_j)$ for each class C_j. $P(C=+)=\frac{3}{5}$ and $P(C=-)=\frac{2}{5}$

Estimate $P\left(\frac{X_j = x_k}{C_j}\right)$ for each value of X_j and each class C_j.

$$P\left(\frac{X_1 = A}{C=+}\right)=\frac{2}{3}$$

$$P\left(\frac{X_1 = B}{C=+}\right)=\frac{1}{3}$$

$$P\left(\frac{X_1 = A}{C=-}\right)=\frac{0}{2}$$

$$P\left(\frac{X_1 = B}{C=-}\right)=\frac{2}{2}$$

Since X_2 is continuous the following conditional probabilities can be defined on X_2

$$P\left(\frac{X_2 = x}{C=+}\right)=g(x;1.73,1.10), \text{where } \mu_{2+}=1.73 \text{ and } \sigma_{2+}=1.10$$

$$P\left(\frac{X_2 = x}{C=-}\right)=g(x;4.45,0.07), \text{where } \mu_{2-}=4.45 \text{ and } \sigma_{2-}=0.07$$

Consider the following dataset for an insurance company as shown in Table 3.4.

TABLE 3.4
Dataset from an Insurance Company

Tid	Refund	Marital Status	Taxable Income	Evade
1	YES	Single	125K	NO
2	NO	Married	100K	NO
3	NO	Single	70K	NO
4	YES	Married	120K	NO
5	NO	Divorced	95K	YES
6	NO	Married	60K	NO
7	YES	Divorced	220K	NO
8	NO	Single	85K	YES
9	NO	Married	75K	NO
10	NO	Single	90K	YES

To determine the conditional probability $P\left(\frac{\text{Income} = 120K}{\text{Evade} = \text{NO}}\right)$ we need to first determine the mean and standard deviation of the samples or tuples highlighted in the above table. The mean and standard deviation, i.e., $\mu_{\text{Income,NO}} = 110$ $\left(\frac{(125+100+70+120+60+220+75)}{7} = 110\right)$ and $\sigma_{\text{Income,NO}} = 54.54$.

Based on the computed mean and standard deviation we can determine

$$P\left(\frac{\text{Income} = 120K}{\text{Evade} = \text{NO}}\right) = \frac{1}{54.54 * \sqrt{2\pi}} e^{-\frac{(120-110)^2}{2(54.54)^2}} = 0.0072$$

Similarly, the conditional probability $\left(\frac{\text{Income} = 120K}{\text{Evade} = \text{YES}}\right) = 1.21517e-9$

Next, we will discuss about the concepts in information theory for predictive analytics.

INFORMATION THEORY FOR PREDICTIVE MODELING

The quantification of information in signals is an interesting field in Information theory. In the context of predictive analytics, some of these concepts are used to characterize or compare probability distributions. The ability to quantify information is used in the decision tree types classifier, to select the variables associated with the maximum information gain. In addition to that the concepts of *entropy* and *cross-entropy* are also important in predictive analytics as they lead to a widely used loss function in classification tasks [1].

The basic intuition behind the information theory is the learning that an unlikely event has occurred is more informative than learning that a likely event has occurred. We define the following concepts in information theory.

Entropy—the measure of impurity in data. If the sample data is completely homogeneous, the entropy is zero, and if the sample is equally divided, it has entropy of one. In a classification task, a variable with a low value for entropy is desired to better segregate the classes. For n classes the entropy is defined as [1]

$$\text{Entropy} = -p_1 * \log_2 p_1 - p_2 * \log_2 p_2 - \ldots - p_n * \log_2 p_n = -\sum_{i=1}^{n} p_i \log_2 p_i$$

Information Gain—the expected reduction in entropy caused by partitioning the instances according to a given attribute. In the context of the decision tree classifier, the information gain is the measure of how much information a feature provides about a class. This measure helps to determine the order of attributes in the nodes of a decision tree [1].

$$\text{information gain} = \text{Entropy}_{\text{parent}} - \text{sum}\left(\text{weighted\%} * \text{Entropy}_{\text{chld}}\right)$$

$$\text{Weighted\%} = \frac{\text{Number of observations in particular child}}{\text{sum}\left(\text{observations in all child node}\right)}$$

Gini impurity—The measure or the probability of misclassifying the instance or an observation [1].

$$\text{Gini} = 1 - \sum_{i=1}^{n} p_i^2$$

where p_i is the probability of an object being classified into a particular class. The degree of Gini index varies from 0 to 1 [1],

- where 0 depicts that all the elements be assigned to a certain class, or only one class exists there.
- the Gini index of value 1 signifies that all the elements are randomly distributed across various classes, and
- a value of 0.5 denotes that the elements are uniformly distributed into some classes.

Consider a simple example to illustrate the concept of entropy and information gain. Let's consider a simple dataset (or sample) that has 1 blue, 2 green, and 3 red balls, stored in a single basket that we cannot see into. Suppose we conduct experiments that consist of randomly drawing one or more balls from the basket.

In this dataset, the entropy is given as

$$\text{Entropy} = -p_b * \log_2 p_b - p_g * \log_2 p_g - p_r * \log_2 p_r$$

Where $p_b = \frac{1}{6}$ is the probability of the blue ball in the dataset, $p_g = \frac{2}{6}$ is the probability of the green ball in the dataset, and $p_r = \frac{3}{6}$ is the probability of the red ball in the dataset

$$\text{Entropy} = -\frac{1}{6} * \log_2 \frac{1}{6} - \frac{2}{6} * \log_2 \frac{2}{6} - \frac{3}{6} * \log_2 \frac{3}{6} = 1.46$$

Now let us consider a scenario where we have 5 blue and 5 green balls, and we have divided them into two baskets where basket 1 consists of 4 blue balls and basket 2 consists of 1 blue ball and 5 green balls.

Before the split, there are 5 blue and 5 green balls. Therefore, the entropy before the split is

$$\text{Entropy}_{\text{before split}} = -\frac{5}{10} * \log_2 \frac{5}{10} - \frac{5}{10} * \log_2 \frac{5}{10} = 1$$

After the split, we have two baskets. In basket 1 with 4 blue balls, the entropy is

$$\text{Entropy}_{\text{basket1}} = -\frac{4}{4} * \log_2 \frac{4}{4} = 0$$

In basket 2 with 1 blue ball and 5 green balls, the entropy is

$$\text{Entropy}_{\text{basket2}} = -\frac{1}{6} * \log_2 \frac{1}{6} - \frac{5}{6} * \log_2 \frac{5}{6} = 0.65$$

Since basket 1 and basket 2 have 4 and 6 elements, respectively, we will assign a weight of 0.4 to basket 1 and 0.6 to basket 2.

Therefore,

$$\text{Entropy}_{\text{after split}} = 04 * 0 + 0.6 * 0.65 = 0.39$$

The information gain can be computed as $1 - 0.39 = 0.61$. The high value for the information gain indicates that the split of the blue and green balls in this scenario resulted in much of the entropy removed.

VIGNETTE The Monty Hall Problem, A Study in Conditional Probability

You may have heard of a television game show called "Let's Make A Deal." In the United States, it started in 1963 and continues to run at this time of writing. It was originally hosted by entertainer Monty Hall, who lends his name to a famous problem in conditional probability based on the show.

At the end of the show, the two top winners get to play the final game for the grand prize. In the game there are three stages blocked by curtains (or doors). Behind one door, there is the grand prize, often a new car. Behind another door, there is a nice prize, but it is not nearly as valuable as the grand prize. Behind a third door, there is a joke prize that is worthless. The contestants do not know what is behind each door, but Monty does know. This is an important fact to remember in the problem formulation.

The game then proceeds along the following lines. The first player (contestant A) is asked to select a door. Then the second player (contestant B) is asked to select from among the remaining two doors. Now Monty prepares to reveal what is behind the doors and also to "Make a deal."

Suppose Monty reveals that contestant B selected the nice (but not the grand prize). After wishing contestant B good luck he now turns to contestant A. Monty offers to let contestant A trade their door choice for the other, unrevealed door. Should contestant A make the deal? What are their odds of winning the grand prize if they pick the other door?

The answer may surprise you. You might say that since the odds of the grand prize being behind any one door was 1/3, that these odds are the same for the door Contestant A already picked and the other door. So their odds of winning the prize if they pick the other door is just 1/2 or 50%. Either the prize is behind there or not. But this reasoning is incorrect.

Contestant A should make the deal because at the outset they had a 1/3 chance of winning the grand prize, but in making the deal and picking the other door they would have a 2/3 chance of winning the grand prize. Why? Because there was always a 1/3 chance of the prize being behind the door Contestant A picked, the door Contestant B picked, and the door no one picked. The odds of the grand prize being behind the door that Contestant A picked is still 1/3. Since we know the grand prize was not behind the door Contestant B picked, there must be a $1 - 1/3 = 2/3$ probability that the grand prize is behind the door not picked. So we make the deal.

If you got the problem wrong, don't be embarrassed. The Monty Hall problem famously fooled a large number of mathematicians at one time. In the September 1990 issue of *Parade Magazine*, a famous advice columnist wrote about the problem. Many professors sent nasty letters to the magazine chastising the author, who was later vindicated, much to the embarrassment of these professors [3].

You can model the Monty Hall problem using Bayes' rule to come to the same conclusion as we did above. Try it!

SUMMARY

This chapter lays the foundations of statistics and probability that are the driving forces behind predictive analytics. Sufficient examples are also provided here to illustrate the major concepts of statistics and probability. It is very essential for the data scientists to master the concepts discussed here, as it will make it much easier for them to follow the materials covered in the future chapters. In addition to that this chapter also discusses about the basics of Naïve Bayes and the information theory that are applicable in designing multiple supervised classifiers.

EXERCISE

Let A be a set of non-negative continuous numeric data
A = {10.1, 12.6, 23.6, 56.7, 67.8, 11.0, 12.45, 16.8, 14.57, 10.1}

1. The Mean, Median, and Mode of set A represented in the format <Mean, Median, Mode> is
   A. <10.1, 12.6, 23.45>
   B. <2.57, 1, 10.1>
   C. <237, 1385, 10>
   D. <23.57, 13.585, 10.1>
   E. <0,0,0>
   Let A be a set of non-negative continuous numeric data
   A = {10.5, 12.8, 23.3, 56.5, 67.4, 11.3, 12.34, 16.78, 14.23, 10.11, 6.78, 14.34}

2. The variance and standard deviation of the subset of A containing the last 5 elements represented in the form of <variance, standard deviation> is
   A. <1, 1>
   B. <15.78, 3.97>
   C. <381.12, 19.52>
   D. <11, 2>
   E. <2.72, 7.4>

3. Is there any linear relationship between the sepal length and sepal width in the Iris dataset? (Clue: Check for the covariance). You can use R or Python script to complete this task. Clearly show the results and provide justification.

4. In the table below, all the **species** in the Iris dataset have been classified as either *small* or *big* (**size**) based on whether the length of the petal is smaller or bigger than the median of all flowers

Size Species	Small	Big
Setosa	1	39
Versicolor	19	11
Virginica	37	3

Determine if there is a relationship between the species and the size. Show all the calculations and provide justification for your conclusion.

Consider the dataset D shown below with the following tuples obtained from the customer database of a large consumer electronics retailer

Rid	Age	Income	Student	Credit Rating	Buys Computer or Not?
1	Youth	High	No	Fair	No
2	Youth	High	No	Excellent	No
3	Middle-aged	High	No	Fair	Yes
4	Senior	Medium	No	Fair	Yes
5	Senior	Low	Yes	Fair	Yes
6	Senior	Low	Yes	Excellent	No
7	Middle-aged	Low	Yes	Excellent	Yes
8	Youth	Medium	No	Fair	No
9	Youth	Low	Yes	Fair	Yes
10	Senior	Medium	Yes	Fair	Yes
11	Youth	Medium	Yes	Excellent	Yes
12	Middle-aged	Medium	No	Excellent	Yes
13	Middle-aged	High	Yes	Fair	Yes
14	Senior	Medium	No	Excellent	No

Based on this dataset, determine if:

5. The customer Jack who is 50 years old, not a university student, earning $140,000 (medium salary range) and with a fair credit history, will buy a computer or not?

6. The customer Simon who is 25 years old, a university student, earning $10,000 (medium salary range) and with an excellent credit history, will buy a computer or not?

7. Compute the entropy of the experiment where one or more balls are drawn randomly from a basket containing 2 green, 5 red, and 6 blue balls?

8. Consider a scenario where 7 blue and 4 green balls have been divided into two bags where bag 1 consists of 4 blue balls and 3 green balls and bag 2 consists of 3 blue balls and 1 green ball. Now compute the following parameters: (a) Entropy before the split, (b) Entropy of bag 1, (c) Entropy of bag 2, and (d) Information gain assuming a weightage of 0.5 for each bags.

NOTES

1 *Kurtosis* refers to the left and right tails of the distribution curve.

2 Informally, independence of two random variables means that the outcome of one does not influence the outcome of another. For example, the outcomes of rolling a die and flipping a coin are independent, regardless of order in which they are done. Conversely, taking the amount of sleep you get on given night and the score you achieve on an exam as random variables, you can argue that one is dependent on another, regardless of which happens first.

REFERENCES

1. Dangeti, P. (2017). *Statistics for Machine Learning*. Packt Publishing Ltd., ISBN 978-1-78829-575-8.

2. Uddin, A., Singh, S., Singh, C.P. "Presentation on Naïve Bayesian Classification", retrieved from http://www.slideshare.net/ashrafmath/naive-bayes-15644818, retrieved on March 4, 2022.
3. Crockett, Z. "The Time Everyone "Corrected" the World's Smartest Woman", *Priceonomics*, retrieved from https://priceonomics.com/the-time-everyone-corrected-the-worlds-smartest/, retrieved on April 2, 2016.

4 Introduction to Machine Learning

The previous chapters focused on providing the necessary background for data analytics. Here, we will discuss some basic machine learning techniques including linear and logistic regression. In addition to that we will also talk about the issues related to bias and variance which plague the machine learning models leading them to overfit. Finally, we will discuss strategies for regularization to address overfitting issues in machine learning model.

VIGNETTE What Is Machine Learning?

Machine learning (ML) is a subset of AI and is the science of training devices or software to perform a task and improve its capabilities by giving it data so it can "learn" over time. The term "machine learning" was coined by artificial intelligence pioneer Arthur Samuel in 1959, while working at IBM. Deep learning (DL) is a subset of machine learning that is designed to function like the human brain using artificial neural networks. We interact daily now with all kinds of ML applications including virtual personal assistants (such as Siri, Alexa, and Bixby), traffic prediction algorithms, purchasing recommenders, credit card fraud detection and much more [1, 2].

Machine learning produces predictive and evolving models based on the data presented. Therefore, conducting assurance of the models is really about verifying and validating the quality of predictions and models. There are two important aspects assurance for both ML and AI in general—explanability and bias [1, 2].

EXPLAINABLE ML/AI –

In AI and ML, *explainability* means that the systems designers can rationalize the system decision making, characterize their strengths and weaknesses, and convey an understanding of how the system will behave in the future. With explainability comes increased trust by an end user to believe and adopt the outcome of the system [1, 2].

BIAS

In ML bias means that the predictive model is somehow producing a result that is unfair to some groups. There are many possible sources bias, including, the use of datasets that are inherently biased; lack of testing; and deployment of technology too soon. Bias can also be caused by statistical anomalies such as overfitting or underfitting or due to skewing or incomplete data in the environment [1, 2].

DOI: 10.1201/9781003278177-4

Bias can be identified and mitigated through a number of means including data collection best practices, analysis of contextual awareness, statistical measures, analysis of variance, outlier detection, causal inference, and many other techniques [1, 2].

STATISTICAL VERSUS MACHINE LEARNING MODELS

Statistical and Machine Learning models are not so distinct from each other. Statistical models are parametric in nature, which means models have parameters on which diagnostics are performed to check the validity of the model. Conversely, the machine learning models are non-parametric in nature, i.e., they do not have any parameters or assumptions. These models can learn by themselves based on the data provided and can produce complex functional models rather than fitting to predefined functions. While statistical models are required to undergo checks for multicollinearity, in machine learning models, the weights of the variables are adjusted automatically to address the multicollinearity problem [1, 2].

Let's start our exploration of ML by discussing certain regression techniques.

REGRESSION TECHNIQUES

Simple linear regression was introduced in Chapter 2. Now we will focus our discussion on Multiple Linear Regression and Multivariate Logistic Regression. First, to recap the regression technique attempts to model the relationship between two types of variables namely the dependent and independent variables, by fitting a linear equation. The most common method for fitting a regression line is the method of least-squares. This method calculates the best-fitting line for the observed data by minimizing the sum of the squares of the vertical deviations from each data point to the line [1, 2].

MULTIPLE LINEAR REGRESSION (MLR) MODEL

A MLR Model describes how a target variable Y relates to two or more X variables. [1, 2]

$$Y = \beta_1 X_1 + \beta_2 X_2 + \beta_3 X_3 + \ldots + \beta_n X_n + \beta_0$$

Using the concept of OLS discussed in Chapter 2 we can determine Sum of Squared Errors (SSE) as:

$$\sum_{i=1}^{N}\left[\left(Y_i - \left(\beta_1 X_{1i} + \beta_2 X_{2i} + \beta_3 X_{3i} + \ldots + \beta_n X_{ni} - \beta_0\right)\right)\right]^2$$

where

- Y_i is an actual value.
- $\hat{Y}_i = \left(\beta_1 X_{1i} + \beta_2 X_{2i} + \beta_3 X_{3i} + \ldots + \beta_n X_{ni} - \beta_0\right)$ is a predicted value.
- $Y_i - \hat{Y}_i$ is a residual error.

Next let's discuss the assumptions for the multiple linear regression.

ASSUMPTIONS OF MLR

MLR has the following assumptions, failing which, the model does not hold true [3]:

1. The dependent variable should be a linear combination of independent variables.
2. There should be no autocorrelation in error terms.
3. The errors (residuals) should have zero mean and should be normally distributed.
4. There should be no or little multi-collinearity.
5. The error terms should be homoscedastic.

Let's present the assumptions of the model in detail and also provide an inventory of tests that are available to test the assumptions [1–3].

The first assumption states that the dependent variable Y should be a linear combination of all the Xs. The presence of the linearity pattern (almost straight red line) in the *Residuals vs. Fitted* plot confirms that this assumption is satisfied (see top left plot in Figure 4.2) [1–3].

To confirm the second assumption, i.e., no autocorrelation in error terms is satisfied, one should perform the Durbin–Watson test (see Sidebar 2). The Durbin–Watson's d tests the null hypothesis that the residuals are not linearly autocorrelated. While d can lie between 0 and 4, if $d \approx 2$ it indicates no autocorrelation, $0 < d < 2$ implies positive autocorrelation, and $2 < d < 4$ indicates negative autocorrelation [1–3].

For confirming the third assumption, i.e., residual errors should have a zero mean and should be normally distributed; the Q-Q plot and also tests such as the Kolmogorov–Smirnov test (see Sidebar 1) will be helpful. For the model to create an unbiased estimate the residual errors should have zero or close to zero mean. On the other hand, if the error terms are not normally distributed, this implies that the confidence intervals will become too wide or narrow, leading to difficulty in estimating the coefficients based on minimization of least squares. In the Q-Q plot if the residuals do not seem to be deviating much compared with the diagonal-like line then it is evident that the errors are normally distributed [1–3].

For confirming the fourth assumption, i.e., the residual errors should be homoscedastic the *Residuals vs. Fitted* plot is observed to make sure that there is no pattern of convergence or divergence. The residual errors should have constant variance with respect to the independent variable. If this assumption is not satisfied, then the performance of the model will degrade. Inconstant variance mostly leads to impractically wide or narrow confidence intervals for the estimates. One reason for not holding homoscedasticity is due to the presence of outliers in the data, which drags the model fit toward them with higher weights [1–3].

Now that we have discussed the assumptions for the fit of the model, we will use an example to illustrate the MLR.

TABLE 4.1
A MLR Fitted on the *mtcars* Dataset

```
input<-mtcars
head(input)
fit_all<-lm(mpg~.,data=input)
summary(fit_all)
```

For performing the MLR we will use the *mtcars* dataset available in R. This dataset contains measurement on 11 different attributes for 32 different cars. In this dataset we are interested in exploring the relationship between a set of variables (independent) and the *miles per gallon* or *MPG* as target(dependent) variable. The *null* and *alternate* hypothesis are [1, 2]

H_0: No linear relationship exists between any of the predictor and the target variable
H_a: There exists at least one linear relationship between the predictor and the target variable

Now consider the R code given in Table 4.1.

Upon fitting the MLR on the *mtcars* dataset we get the output show in Figure 4.1.

From Figure 4.1 it is evident that none of the *p-values* is significant for the predictors. However, the *p-value* of the model $(p = 3.79e-07)$ suggest rejecting the null hypothesis and concluding that there exists at least one linear relationship between the target and the predictor variables. The adjusted R-squared of 0.8066 also suggest that there is a strong relationship between the predictors and target variable. Now let's check to see if the assumptions of MLR are met for this model [2, 3].

To check if the mean of the residuals is zero the following R code can be used

```
mean(fit_all$residuals)
```

This code results in a mean value of $7.45e-17$, which is approximately zero. Therefore, it is confirmed that the mean of the residuals (the residual error) is zero.

To check for the evidence of homoscedasticity plot the *Residual vs. Fitted* plot (see Figure 4.2) using the following command

```
par(mfrow=c(2,2)) # set 2 rows and 2 column plot layout
plot(fit_all)
```

Since the red line in the top left plot (see Figure 4.2) is not a flat or a straight line, there is no evidence of homoscedasticity. The residuals are first decreasing and then increasing. Now see the Normal Q-Q plot, the points lie exactly on the line. Therefore, the residuals are normally distributed. However, it is okay to expect some deviation of the points particularly near the ends (see Normal Q-Q plot in Figure 4.2).

Now let's check for the autocorrelation of the residual errors. In order to perform the Durbin–Watson test (see sidebar 2) execute the R code shown in Figure 4.3.

```
lmtest::dwtest(fit_all)
```

```
Call:
lm(formula = mpg ~ ., data = input)

Residuals:
    Min      1Q  Median      3Q     Max
-3.4506 -1.6044 -0.1196  1.2193  4.6271

Coefficients:
            Estimate Std. Error t value Pr(>|t|)
(Intercept) 12.30337   18.71788   0.657   0.5181
cyl         -0.11144    1.04502  -0.107   0.9161
disp         0.01334    0.01786   0.747   0.4635
hp          -0.02148    0.02177  -0.987   0.3350
drat         0.78711    1.63537   0.481   0.6353
wt          -3.71530    1.89441  -1.961   0.0633 .
qsec         0.82104    0.73084   1.123   0.2739
vs           0.31776    2.10451   0.151   0.8814
am           2.52023    2.05665   1.225   0.2340
gear         0.65541    1.49326   0.439   0.6652
carb        -0.19942    0.82875  -0.241   0.8122
---
Signif. codes:  0 '***' 0.001 '**' 0.01 '*' 0.05 '.' 0.1 ' ' 1

Residual standard error: 2.65 on 21 degrees of freedom
Multiple R-squared:  0.869,    Adjusted R-squared:  0.8066
F-statistic: 13.93 on 10 and 21 DF,  p-value: 3.793e-07
```

FIGURE 4.1 MLR model fitted on the *mtcars* dataset.

The Durbin-Watson test suggests that there is a positive autocorrelation as $d < 2$. Therefore, the second assumption of no autocorrelation in the residual errors is not satisfied.

Finally, to check for the assumption of no multicollinearity among predictors, the Variance Inflation Factor (VIF) value is obtained for each predictor. For a good fit of the model, it is desired to have the VIF value of all the predictors to be under 4. The VIF value for the predictors can be obtained using the R code shown in Figure 4.4.

```
library(car)
vif(fit_all)
```

Here, we see that several predictors have VIF value greater than 4. If the VIF of the predictors is greater than 4, then remove those predictor attributes and fit the MLR again with the remaining predictor attributes. To begin, start by removing the predictor attribute that has the highest VIF value. Each time a new MLR model is fitted, check for the VIF of each predictor. This step has to be continued until all the remaining predictors of the MLR model have a VIF value less than 4.

To fit an MLR model on a dataset you can start with two predictors in the beginning and then keep on adding more predictors to the model. This process is known as *forward selection*. Alternatively, one can start with all the predictors and then keep on excluding predictors one at a time from the model. This process is known as *backward selection*. Both together are collectively known as *stepwise regression*. However,

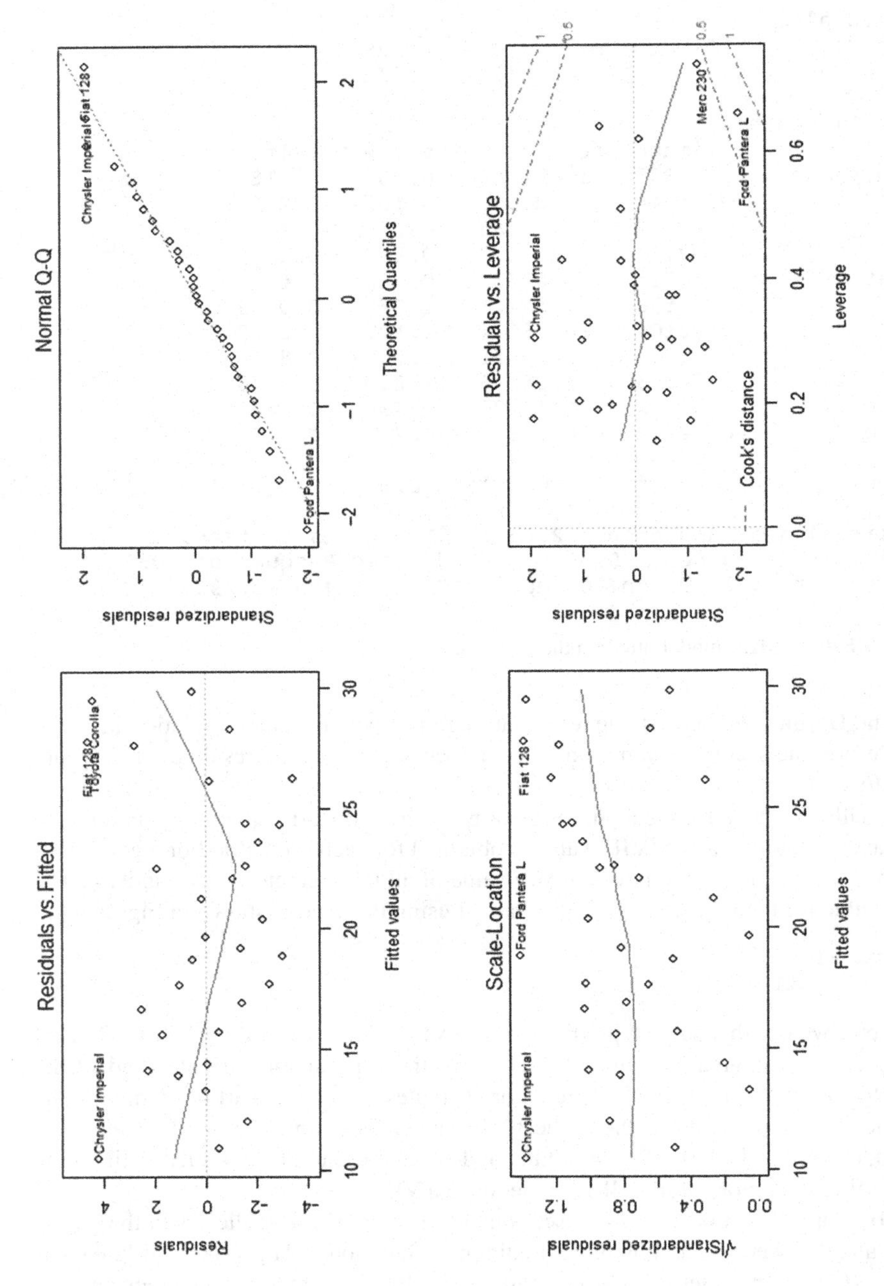

FIGURE 4.2 Plots for the MLR model fitted on the *mtcars* dataset.

```
        Durbin-Watson test

data:   fit_all
DW = 1.8609, p-value = 0.1574
alternative hypothesis: true autocorrelation is greater than 0
```

FIGURE 4.3 Output of the Durbin–Watson test on the MLR model.

```
      cyl      disp        hp      drat        wt      qsec        vs        am
15.373833 21.620241  9.832037  3.374620 15.164887  7.527958  4.965873  4.648487
     gear      carb
 5.357452  7.908747
```

FIGURE 4.4 Variance inflation factor test for multicollinearity on the MLR model.

each time a model is fitted the assumptions have to be to be verified to make sure that they are all satisfied [2].

Next let's look at the Multivariate Logistic Regression (MLogR) model.

INTRODUCTION TO MULTINOMIAL LOGISTIC REGRESSION (MLOGR)

Before we discuss the *Multinomial Logistic Regression (MLogR)*, let's consider Logistic Regression (LogR) in general. In simple terms, LogR's are linear models for binary outcomes. There are several events for which we often observe a yes/no outcome. For example, did the customer like the product? Did they take a test drive of the latest model of electric vehicle? Did the customer renew the subscription? And so on. All these kinds of questions have outcomes that are binary in nature, i.e., either a "yes" or "no."[1] Though it is possible to fit such a model with a typical linear regression model, however the most efficient and useful way to fit such outcomes is with a LogR model also known as a logit model [1, 2, 4].

The core concept behind the LogR model is that it relates the *probability* of an outcome to an *exponential* function of a predictor variable. By modeling the *probability* of an outcome, a LogR model accomplishes two things. First, it directly models the probability or the likelihood that an event will occur and secondly, it limits the model to the appropriate range for the proportion, which is [0,1] [1, 2, 4].

The equation for the LogR function is [1]

$$\text{LogR: } p(y) = \frac{e^{v_x}}{e^{v_x} + 1}$$

Here the outcome of interest is $y \in [0,1]$ (target variable) and we compute the likelihood $p(y)$ as a function of v_x. v_x can take any real value, so we are able to treat it as a continuous function in a linear model. In that case v_x indicates the importance of the corresponding predictor variable. The likelihood of y or the *logistic* value is less than 50% when v_x is negative, is 50% when $v_x = 0$, and is above 50% when v_x is positive.

Let's consider some examples to illustrate the logistic value (y) or the *plogis*(). The following commands are in R [1]

```
exp(0) / (exp(0) + 1) # equivalent to plogis()
0.5
plogis(-Inf) # infinite dispreference = likelihood 0
0
plogis(2) # moderate preference = 88% chance (e.g., of
  purchase)
0.8807971
plogis(-0.2) # weak dispreference
0.450166
```

On the other hand, a *logit* model determines the value of v_x from the logarithm of the relative probability of occurrence of y[1]:

$$\text{logit}: v_x = \log\left(\frac{p(y)}{1-p(y)}\right)$$

To compute the *logit* value (v_x), R has an in-built function *qlogis*() as shown below

```
log(0.5/(1-0.5)) # indifference = 50% likelihood = 0 utility
0
log(0.88/(1-0.88)) # moderate high likelihood
1.99243
qlogis(0.88) # equivalent to hand computation
1.99243
```

Now let's discuss about the MLogR. MLogR is used to model the nominal outcome (target) variable as a linear combination of the predictor variables. The predictor variables can be either dichotomous (i.e., binary) or continuous (i.e., interval or ratio in scale) in nature. It is a simple extension of LR that allows for more than two categories of the dependent variable. The MLogR uses the maximum likelihood estimation to determine the probability of the target variable. Before performing the MLogR, it is important to carefully consider the sample size (a minimum of 10 cases per predictor variable) requirements and the examination of the outlying cases in addition to performing the univariate, bivariate, and multivariate assessment [1]. More importantly, multicollinearity should be evaluated with simple correlations among the predictors. Multivariate diagnostics can be used to access the multivariate outliers and for exclusion of outliers. MLogR does not assume normality, linearity, or homoscedasticity. The following are the assumptions of the MLogR [1, 4]

- The target variable should be measured at the nominal level with more than or equal to three values.

- One or more predictors are continuous, ordinal, or nominal (including dichotomous variables). However, ordinal predictors must be treated as being either continuous or categorical.
- The observations are independent, and the target variable should have mutually exclusive and exhaustive categories.
- There should be no multicollinearity. Multicollinearity occurs when two or more predictors are highly correlated with each other.
- There needs to be a linear relationship between any continuous predictors and the logit transformation of the target variables.
- There should be no outliers, high leverage values, or highly influential points for the scale/continuous variables.

Now let's fit a MLogR model to a dataset. In R, MLogR model is fit using the *generalized linear model* (GLM) using the *glm*() function, which can handle target variables even if they are not normally distributed. GLM models can also relate normally or non-normally distributed predictors to a non-normal outcome using a function known as a *link*. This means that GLM is a single, consistent framework that can fit models for many different distributions. The *glm* function takes an argument *family = binomial* that specifies the distribution for the target variable. The default link function for a binomial model is the logit function [1, 2].

To demonstrate the MLogR model fit we consider the amusement park dataset. In this dataset we have data on the sales of season tickets to the park. The target variable is the season ticket *pass sales* (with values of *yes* or *no*), based on two predictors namely the *channel* used to extend the offer (email, postal mail, or in-person at the park) and *promotion criteria*, i.e., whether the seasonal tickets were sold as bundles with promotional offers (free parking) or not. The MLogR model can be used to address queries such as: Are customers more likely to purchase the season pass when it is offered in the bundle (with free parking), or not?

The R code for fitting the MLogR model is given in Figure 4.5. The amusement park dataset can be obtained from the link http://goo.gl/J8MH6A [1]

```
# Retrieve the amusement park dataset
data <- read.csv("http://goo.gl/J8MH6A")
data$Pass <- as.factor(data$Pass)
data$Channel <- as.factor(data$Channel)
data$Promo <- as.factor(data$Promo)
summary(data)
```

Figure 4.5 describes the composition of the amusement park dataset. The *channel* or the mode of selling the seasonal passes are either through *email*, *mail* or *park*.

```
   Channel          Promo            Pass
 Email: 633   Bundle  :1674   NoPass :1567
 Mail :1328   NoBundle:1482   YesPass:1589
 Park :1195
```

FIGURE 4.5 Composition of the amusement park dataset.

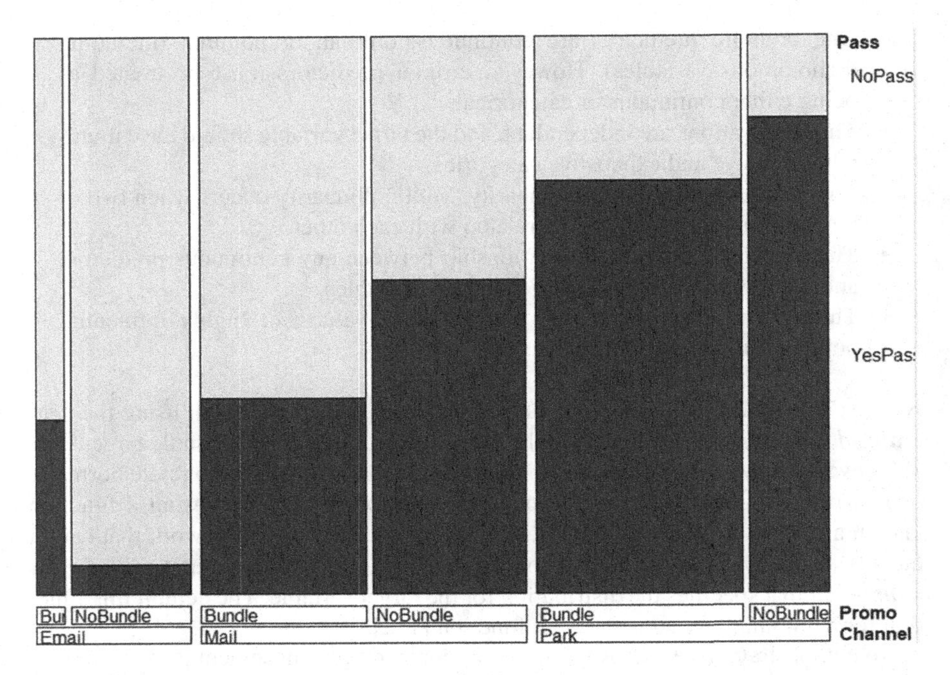

FIGURE 4.6 Mosaic plot demonstrating the sales of seasonal passes through different channels. (Source: Figure adapted from Chapman, C., and Feit, E. McDonnell, "R for Marketing Research and Analytics," *Springer*, ISBN 978-3-319-14436-8, 2015.)

The *promotion* is either through *bundle* or *nobundle*. This dataset has a total of 3156 rows and 3 columns. The *channel* that was most successful in selling seasonal passes was the park, regardless of whether the promotion was offered or not (see Figure 4.6). A good way to visualize this data is with *mosaic* plots. Using the *vcd* package a *doubledecker* plot can be constructed as shown in Figure 4.6.

```
Install.packages("vcd")
library(vcd)
doubledecker(table(data))
```

The sales of seasonal passes are very successful at the park and are very unsuccessful by email. Therefore, we have to consider both the *channel* and the *promotion* as predictors for determining the outcome of the *sale of pass*.

```
mod <- glm(Pass ~ Promo + Channel + Promo:Channel, data=data,
  family=binomial)
summary(mod)
exp(confint(mod))
```

Here it is important to remember that we also need to consider the interaction effect of the predictors *channel* and *promotion*. This interaction effect is considered as a third predictor in the model indicated as (Promo:Channel). Except for the predictor

```
Call:
glm(formula = Pass ~ Promo + Channel + Promo:Channel, family = binomial,
    data = data)

Deviance Residuals:
    Min       1Q   Median       3Q      Max
-1.9577  -0.9286   0.5642   0.7738   2.4259

Coefficients:
                         Estimate Std. Error z value Pr(>|z|)
(Intercept)               -0.7813     0.1959  -3.989 6.64e-05 ***
PromoNoBundle             -2.1071     0.2783  -7.571 3.71e-14 ***
ChannelMail                0.1632     0.2115   0.772     0.44
ChannelPark                1.8340     0.2107   8.702  < 2e-16 ***
PromoNoBundle:ChannelMail  2.9808     0.3003   9.925  < 2e-16 ***
PromoNoBundle:ChannelPark  2.8115     0.3278   8.577  < 2e-16 ***
---
Signif. codes:  0 '***' 0.001 '**' 0.01 '*' 0.05 '.' 0.1 ' ' 1

(Dispersion parameter for binomial family taken to be 1)

    Null deviance: 4375.0  on 3155  degrees of freedom
Residual deviance: 3393.5  on 3150  degrees of freedom
AIC: 3405.5

Number of Fisher Scoring iterations: 5

          Waiting for profiling to be done...
                                          2.5 %     97.5 %
          (Intercept)                0.30864489  0.6667743
          PromoNoBundle              0.06985845  0.2087813
          ChannelMail                0.78297153  1.7975584
          ChannelPark                4.17076910  9.5473418
          PromoNoBundle:ChannelMail 10.98615152 35.7670765
          PromoNoBundle:ChannelPark  8.80206901 31.8934797
```

FIGURE 4.7 MLogR model on the amusement park dataset.

ChannelMail all the predictors have a *p-value* that is significant (see Figure 4.7). Also note here that the interaction of *promotion* with *channel* is statistically significant for the mail and in-park channels, as opposed to the email channel. In the odds ratios, we see that the promotion without a bundle is 10–35% as effective through the mail and is 8%–31% as effective through in-park channels as it is in email. Therefore, we can now conclude that the promotion without the bundle to be effective depends on the promotion channel. There is good reason to continue the promotion campaign by email, but its success there does not necessarily imply success at the park or through a regular mail campaign (see Figure 4.7) [1].

Next, we will discuss the *bias*, *variance*, and the trade-off between the promotion options. Finally, we'll explore issues related to overfitting in linear regression models and how the regularization techniques can be used to obtain an optimal model.

Bias versus Variance Trade-off

Every analytical (predictive) model has both *bias* and *variance* error components in addition to *white noise*.[2] Bias and variance are inversely related to each other. This means that in an effort to reduce one error component in the model, the other error component of the model will increase. The true art lies in creating a good fit by balancing both the errors. The ideal model should have both low bias and low variance. This is where the trade-off becomes necessary [2].

The errors from the bias component come from erroneous assumptions in the underlying learning algorithm. Actually, high bias can cause an algorithm to miss the relevant relationship between the predictors and the target variable resulting in underfitting. On the other hand, the errors from the variance component come from the sensitivity to change in the fit of the model, i.e., even a small change in the training data can result in high variance, also well-known as an overfitting problem [2].

Both the LR/MLR and MLogR are good examples of a high bias model. In both modeling techniques, the fit of the model is merely a straight line and may have a high error component. This is because a linear model cannot approximate the underlying data well. On the other hand, an example of a high variance model is a decision tree in which the model may create too much wiggly curve as a fit. Models created using the decision tree would result in drastic changes in the fit of the curve which a small change takes place in the training data [2].

In order to address the trade-off between bias and variance, the state-of-the-art models use high variance models such as decision trees and performing ensemble on top of them. This results in reducing the errors caused by high variance without compromising the increase in errors due to the bias component [2]. The best example of learning algorithm or classifier in this category is Random Forest, in which many decision trees are grown independently, and ensemble is performed to come up with the best fit model. The decision tree and the Random Forest classifier will be discussed in detail in Chapter 6.

Overfitting and Underfitting

Earlier we mentioned overfitting and underfitting. Generally, overfitting occurs when a very complex statistical model or learning algorithm suits the observed data because it has too many predictors compared to the number of observations. The problem with overfitting is that an incorrect model can perfectly fit data, just because it is quite complex compared to the amount of data available. Again, it is also possible for overfitting to occur when the amount of data is adequate. As a result, when the obtained model is used to predict new observations, mispredictions occur, because it is not able to generalize [1, 2].

Now we'll discuss overfitting and underfitting of the model in the context of the regression. The concept of overfitting is very important in regression analysis. Usually, a learning algorithm is trained using a set of examples (the training set), the output of which is already known. It is assumed here that the learning algorithm that is generalized will reach a state in which it will be able to predict new instances that it has not yet seen. On the contrary, underfitting occurs when a regression algorithm cannot capture the underlying trend within the data. Underfitting would occur, for example, when fitting a linear model to some nonlinear data. Both the overfitting and underfitting models would suffer from poor predictive performance [1, 2].

Now let's turn our discussion to the topic of regularization.

Regularization

To select relevant features for the model it is possible to adopt methods that use all the predictors but adjust the coefficients of the predictors by bringing them to a very small close to zero or exact zero values also referred to as shrinkage. These methods are also referred to as *automatic feature selection* methods, as they tend to improve generalization. They are called *regularization methods* [2]. When there are many variables available, the least square estimate of a linear model often results in low bias but high variance with respect to models with fewer variables. Under these conditions there is an overfitting problem. To improve precision prediction with small variance but with greater bias we can use variable selection methods and dimensionality reduction techniques. However, these methods may be unattractive due to computational burdens in the first case or provide a difficult interpretation in the other case [2, 5].

Another way to address the problem of overfitting is to modify the estimation method by neglecting the requirement of an unbiased parameter estimator and consider using a biased estimator, which may have smaller variance. There are several biased estimators mostly based on regularization namely *Ridge*, *Lasso*, and *ElasticNet* regressors. Regularization is a technique where the regression coefficients are shrunk by introducing some type of penalty. In this chapter, we will discuss about the Ridge and Lasso regression [2, 5].

There are two types of regularization, L1 and L2. In L1 regularization or Lasso or L1 norm we shrink the parameter to zero. Sparse L1 norm are created when input features have weights closer to zero. In sparse solution majority of the features have zero weights and a very few numbers of features have non-zero weight. Thus, L1 regularization results in feature selection. The undesirable input features are assigned a weight of zero and useful features are assigned a non-zero weight. In L1 regularization, the absolute value of the weights is penalized. Therefore, Lasso produces a model that is simple, interpretable, and contains a subset of the total input features [2, 5].

Conversely, L2 regularization or Ridge regression the regularization term is the sum of square of all the feature weights. L2 regularization forces the weights of the predictors to be small but not zero. Thus, the solution is non-sparse. L2 is not robust to outliers. This is because the square terms blow up the error differences of the outliers. However, the regularization term tries to fix this by penalizing the weights. Ridge regression performs better when all the input features are instrumental to the target variable. The weights of the predictors are roughly of equal size. **Elastic net regularization** on the other hand is a combination of both L1 and L2 regularization [2, 5].

Let's summarize the differences between the L1 and L2 regularization techniques [2, 5]:

1. L1 penalizes the sum of absolute value of weights and L2 penalizes the sum of square weights.
2. L1 has a sparse solution and L2 has a non-sparse solution.
3. L1 has multiple solutions but L2 has one solution.

4. L1 has built in feature selection and L2 has no feature selection.
5. L1 is robust to outliers but L2 is not robust to outliers.
6. L1 generates models that are simple and interpretable but cannot learn complex patterns. On the other-hand L2 gives better prediction when the target variable is a function of all the predictors.
7. L2 can learn complex data patterns from the data but L1 is not so good at it.

First, we'll discuss about the L2 norm or the Ridge regression.

Ridge Regression

Consider a scenario where there is an issue with multicollinearity. When multicollinearity occurs, the least square estimates are unbiased, but their variances are too large, so they may be far from the true value. By adding a degree of bias to the regression estimates the Ridge regression reduces the standard errors. Ridge regression is very similar to least squares. The Ridge regression coefficients β are the values that minimize the following expression [2, 5]:

$$\sum_{i=1}^{n}\left(y_i - \beta_1 x_i + \beta_0\right)^2 + \lambda * \beta_1{}^2 = RSS + \lambda * \beta_1{}^2$$

Here, $\lambda \geq 0$ is a tuning parameter which needs to be determined. The term $\lambda * \beta_1{}^2$ is a shrinkage penalty that decreases when the β parameters shrink towards zero. Parameter λ controls the relative impact of the two components: RSS and the penalty term. If $\lambda = 0$, the Ridge regression coincides with the least square's method. When $\lambda \to \infty$, all estimated coefficients tend to zero [5].

Ridge regression produces different estimates for different values of λ. Therefore, determining the optimal choice of λ is crucial and is usually done with a technique called *cross-validation*. Note that the shrinkage penalty is not applied to β_0, as it would not make sense. Ridge regression addresses the problem by estimating regression coefficients using the following equation [2, 5]

$$\beta = \left(X^T * X + \lambda * I\right)^{-1} * X^T * Y$$

Here, λ is the Ridge parameter and I is the *identity matrix*.[3] Small positive values of λ are desirable to reduce the variance of the estimates. While biased, the reduced variance of Ridge estimates often result in a smaller mean square error. The matrix $\left(X^T * X + \lambda * I\right)^{-1}$ is not singular. In the Ridge regression it is advisable to standardize all predictors before estimating the model. To standardize the predictors, the procedure to be followed is to subtract their means and divide by their standard deviations [2, 5].

Here we show using a R code how Ridge regression can handle the multicollinearity issue in the *Seatbelts* dataset. First we need to install the *glmnet* package [5] to explore the dataset. From Figure 4.8 it is evident that several variables are highly correlated.

	drivers	front	rear	kms	PetrolPrice	VanKilled	law
drivers	1.00	0.81	0.34	-0.44	-0.46	0.49	-0.45
front	0.81	1.00	0.62	-0.36	-0.54	0.47	-0.56
rear	0.34	0.62	1.00	0.33	-0.13	0.12	0.03
kms	-0.44	-0.36	0.33	1.00	0.38	-0.50	0.49
PetrolPrice	-0.46	-0.54	-0.13	0.38	1.00	-0.29	0.39
VanKilled	0.49	0.47	0.12	-0.50	-0.29	1.00	-0.39
law	-0.45	-0.56	0.03	0.49	0.39	-0.39	1.00

FIGURE 4.8 Correlation between predictors in the *Seatbelts* dataset.

```
library (car) # Load the "car" library. This library will help
  to compute the VIF
#Install and load the glmnet package
install.packages("glmnet")
library (glmnet)
# Lod the dataset "Seatbelts"
data(Seatbelts, package="datasets") # initialize data
inputData <- data.frame (Seatbelts) # Duplicate the dataset
colnames(inputData)[1] <- "response" # rename "Driverskilled"
  to response
XVars <- inputData[, -1] # Obtain all the predictors
round(cor(XVars), 2) # Correlation Test
```

To begin, the dataset is split into train and test dataset in the ratio of 80% and 20%, respectively.

```
set.seed(100) # set seed to replicate results
# Construct the training and test dataset
trainingIndex <- sample(1:nrow(inputData),
  0.8*nrow(inputData)) # indices for 80% training data
trainingData <- inputData[trainingIndex, ] # training data
testData <- inputData[-trainingIndex, ] # test data
```

Now we'll perform Ridge regression by executing the R code below. Note the values assigned to the hyperparameters namely the *nlambda* and *lambda.min.ratio* in the *glmnet* function. The hyperparameter *alpha* in the *glmnet* function is set to 0 (zero) for performing the Ridge regression [5].

```
x <- model.matrix(response~., trainingData)[,-c(1,9)]
y <- trainingData$response
# The term [,-c(1,9)] is used to remove the intercept. If
  alpha=0, then Ridge regression is used
# nlambda=100: Set the number of lambda values (the default is
  100)
# lambda.min.ratio=0.0001: Set the smallest value for lambda,
  as a fraction of lambda.max, the # entry value (that is, the
  smallest value for which all coefficients are zero)
RidgeMod <- glmnet(x,y, alpha=0, nlambda=100, lambda.min.
  ratio=0.0001)
```

```
plot(RidgeMod,xvar="lambda",label=TRUE)
CvRidgeMod <- cv.glmnet(x, y, alpha=0, nlambda=100,lambda.min.
  ratio=0.0001)
plot(CvRidgeMod)
```

From Figure 4.9, it is evident that when *lambda* is very large (e.g., the log of lambda is ten), the regularization effect dominates the squared loss function, and the coefficients tend to zero. At the beginning of the path, as lambda tends toward zero and the solution tends toward the Ordinary Least Square (OLS) the coefficients exhibit big oscillations (because they are unregularized). In practice, it is necessary to tune *lambda* in such a way that a balance is maintained between both [5].

Figure 4.10 shows the cross-validation curve and the upper and lower standard deviation curves. In the beginning the Mean Squared Error (MSE) is very high and then at some point it kind of levels off. This seems to indicate that the model is doing well. There are two vertical lines: one is at the minimum, and the other vertical line is within one standard error of the minimum. The second line is a slightly more restricted model that does almost as well as the minimum. The two lambda values indicate [5]:

- lambda.min is the value of λ that gives the minimum mean cross-validated error
- lambda.1se, gives the most regularized model such that the error is within one standard error of the minimum

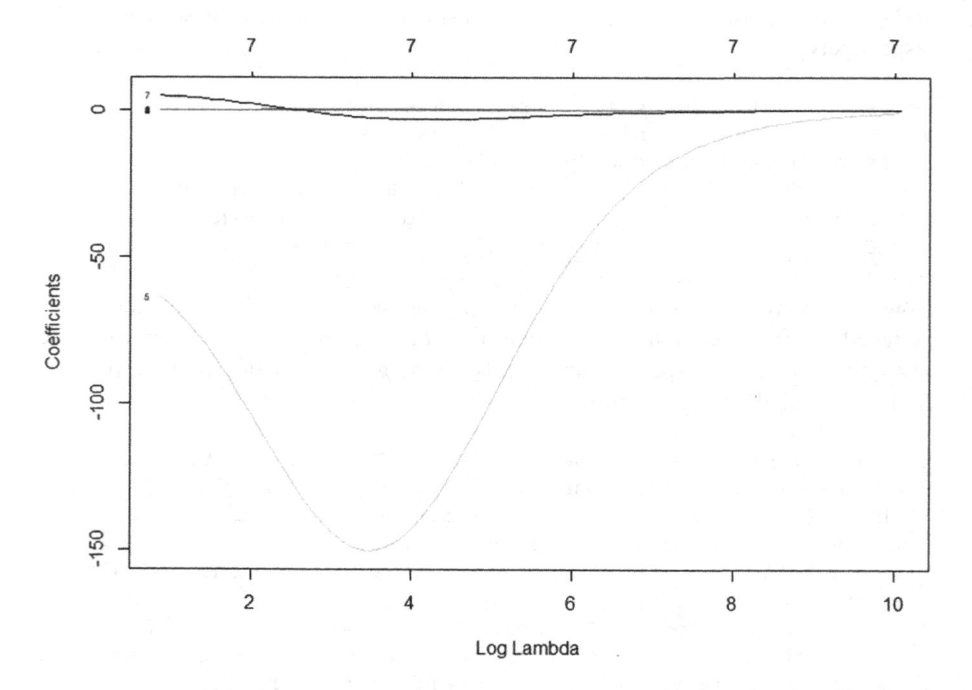

FIGURE 4.9 Coefficients of the predictors with respect to *Log Lambda*. (Source: Figure adapted from Hastie, T., Qian, J., Tay, K. (2021). An Introduction to glmnet. Retrieved from https://glmnet.stanford.edu/articles/glmnet.html, retrieved on April 11, 2022.)

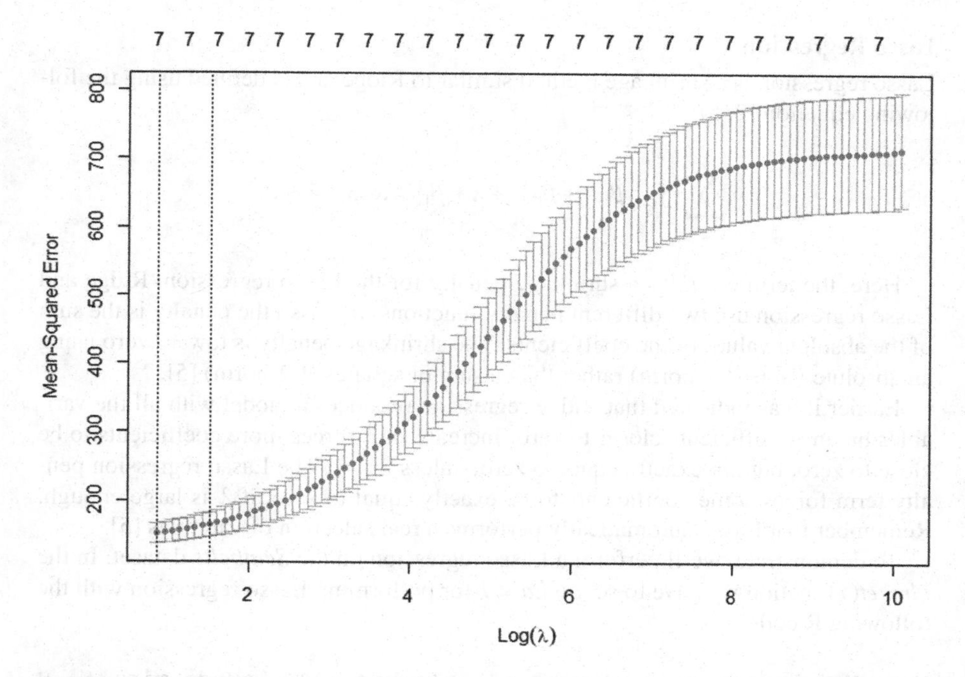

FIGURE 4.10 MSE with respect to *Log Lambda*. (Source: Figure adapted from Hastie, T., Qian, J., Tay, K. (2021). An Introduction to glmnet. Retrieved from https://glmnet.stanford.edu/articles/glmnet.html, retrieved on April 11, 2022.)

There are all seven variables (see top of the plot indicating the number of nonzero variables) in the model (six variables, plus the intercept), and no coefficient is zero [5].

```
Now we'll extract the value of best lambda.

best.lambda <- CvRidgeMod$lambda.min
best.lambda # print the best lambda value
# Now determine the coefficient of the nonzero predictors
predict(RidgeMod, s=best.lambda, type="coefficients")[1:7, ]
```

The best value of the lambda should be ~2.4. Figure 4.11 lists the coefficients of the nonzero predictors. It can be observed that all the six predictors have very small coefficients but are not zero [5].

Next, we'll discuss the L1 norm or Lasso regression.

```
  (Intercept)       drivers         front          rear           kms   PetrolPrice
-1.632769e+00  6.679323e-02  1.753720e-02  5.102092e-04  2.327151e-04 -6.377056e+01
    VanKilled
 8.731971e-03
```

FIGURE 4.11 Coefficients of the nonzero predictors of the model.

Lasso Regression

Lasso regression is a shrinkage method similar to Ridge and is defined using the following equation [2, 5]:

$$\sum_{i=1}^{n}\left(y_i - \beta_1 x_i + \beta_0\right)^2 + \lambda * |\beta_1| = RSS + \lambda * |\beta_1|$$

Here, the term $\lambda * |\beta_1|$ is a shrinkage penalty for the Lasso regression. Ridge and Lasso regression use two different penalty functions. In Lasso the penalty is the sum of the absolute values of the coefficients. The shrinkage penalty is toward zero using an absolute value (L1-norm) rather than a sum of squares (L2-norm) [5].

Earlier it was indicated that Ridge regression produces a model with all the variables having coefficients closer to zero. Increase in λ forces more coefficients to be close to zero, but not exactly equal to zero, unless $\lambda = \infty$. The Lasso regression penalty term forces some coefficients to be exactly equal to zero, if λ is large enough. Remember that Lasso automatically performs a real selection of variables [5].

To demonstrate, we'll perform a Lasso regression on the *Seatbelts* dataset. In the *glmnet*() function we have to set *alpha = 1* for performing Lasso regression with the following R code.

```
LassoMod <- glmnet(x,y, alpha=1, nlambda=100, lambda.min.
  ratio=0.0001)
plot(LassoMod,xvar="norm",label=TRUE)
CvLassoMod <- cv.glmnet(x, y, alpha=1, nlambda=100,lambda.min.
  ratio=0.0001)
plot(CvLassoMod)
```

In resulting plot, shown in Figure 4.12, each curve corresponds to a variable.

This plot shows the path of its coefficient against the L1-norm of the coefficient when λ is varying. The axis indicates the number of nonzero coefficients at the current λ, which is the effective degrees of freedom for the Lasso [5].

Figure 4.13 includes the cross-validation curve (red dotted line), and upper and lower standard deviation curves along the λ sequence (error bars). In the beginning the MSE is very high, and the coefficients are restricted to be too small, and then at some point, it levels off. This seems to indicate that the model is doing well. The lambda.1se (second line to the left) is the one which is the largest λ value, within one standard error of λ min. The second line is slightly more restricted indicating that the model is almost as well as the minimum [5].

Now let's extract the value of best lambda with the following code [5].

```
best.lambda <- CvLassoMod$lambda.min
best.lambda
coef(CvLassoMod, s = "lambda.min")
```

The best value of the lambda should be ~ 0.27. Figure 4.14 lists the coefficients of the nonzero predictors. We can see that the Lasso method is able to select variables.

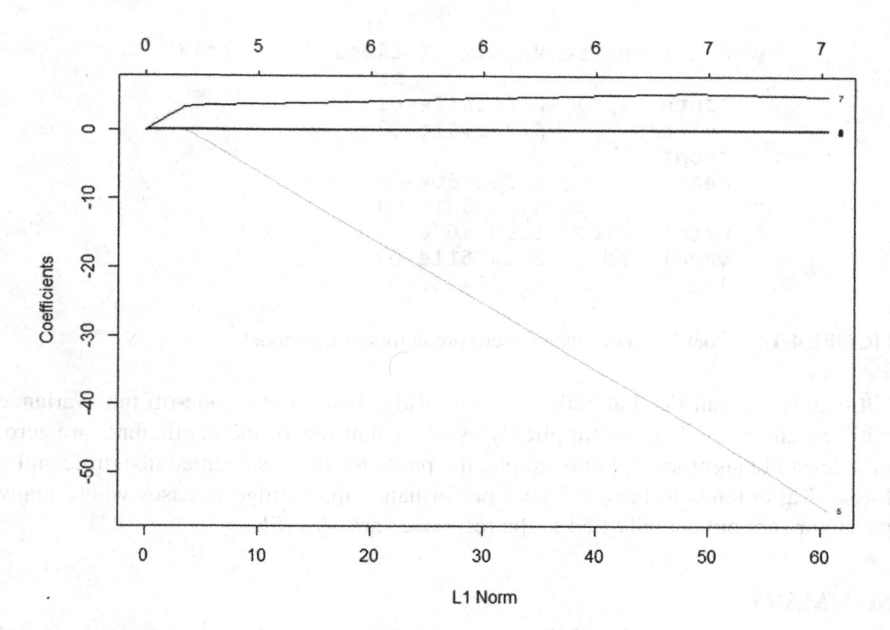

FIGURE 4.12 Coefficients of the predictors with respect to L1 norm.

(Source: Figure adapted from Hastie, T., Qian, J., Tay, K. (2021). An Introduction to glmnet. Retrieved from https://glmnet.stanford.edu/articles/glmnet.html, retrieved on April 11, 2022.)

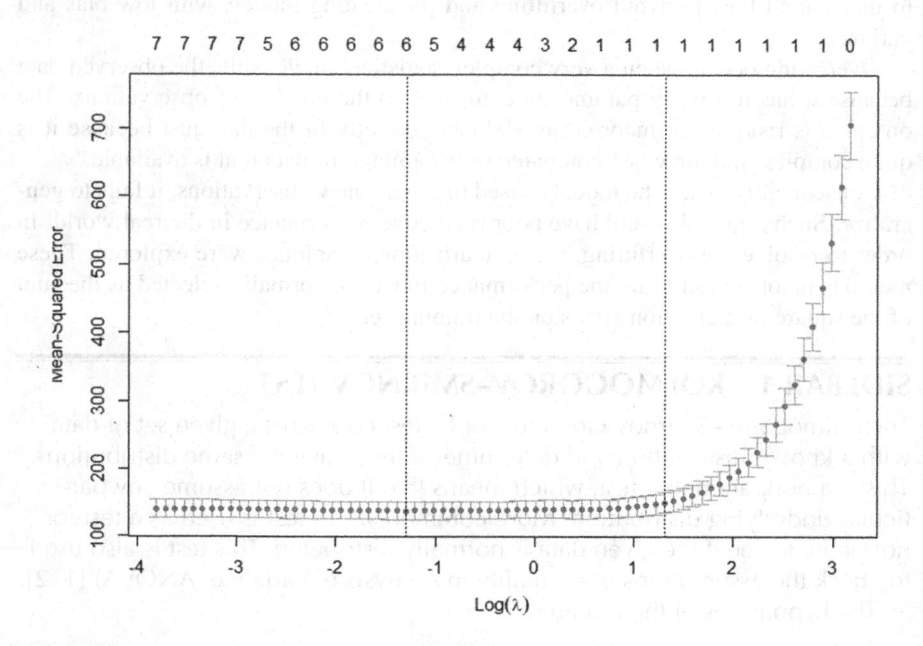

FIGURE 4.13 MSE with respect to *Log Lambda*.

(Source: Figure adapted from Hastie, T., Qian, J., Tay, K. (2021). An Introduction to glmnet. Retrieved from https://glmnet.stanford.edu/articles/glmnet.html, retrieved on April 11, 2022.)

```
8 x 1 sparse Matrix of class "dgCMatrix"
                        s1
(Intercept) -2.001935e+01
drivers      8.231941e-02
front        .
rear         6.282480e-04
kms          4.231001e-04
PetrolPrice -1.990867e+01
VanKilled   -2.347611e-02
law          4.372455e+00
```

FIGURE 4.14 Coefficients of the nonzero predictors of the model.

Ultimately, we can say that both Lasso and Ridge balance the trade-off bias-variance with the choice of λ. Lasso implicitly assumes that few of the coefficients are zero, or at least not significant. For example, the predictor *front* is deemed insignificant by Lasso. Lasso tends to have a higher performance than Ridge in cases where many predictors are not actually tied to the response variables [5].

SUMMARY

In this chapter, we have learned about multiple liner regression, logistic regression, and regularization techniques. In addition to that we have learnt how to achieve generalization for our models. Both the Ridge and Lasso techniques were explored here to understand how to avoid overfitting and for creating models with low bias and variance.

Overfitting occurs when a very complex statistical model suits the observed data because it has too many parameters compared to the number of observations. The outcome is risky as an incorrect model can perfectly fit the data just because it is quite complex in logic when compared to the amount of data that is available.

Consequently, when the model is used to predict new observations, it fails to generalize. Such a model would have poor predictive performance in the real world. In order to resolve the overfitting, the regularization techniques were explored. These methods involve modifying the performance function, normally selected as the sum of the square of regression errors on the training set.

SIDEBAR 1 KOLMOGOROV–SMIRNOV TEST

The Kolmogorov–Smirnov Goodness of Fit test compares a given set of data with a known distribution and determines if they have the same distribution. This is a nonparametric test, which means that it does not assume any particular underlying distribution. More commonly, this test is used as a test for normality to see if the given data is normally distributed. This test is also used to check the assumptions of normality in Analysis of Variance (ANOVA) [1, 2].

The hypotheses of the test are:

H_0: The data comes from the specified distribution

H_a: At least one value does not match the specified distribution

If the null hypothesis is not rejected, then it can be confirmed that both the data have the same distribution.

The Kolmogorov–Smirnov test statistic measures the largest distance between the empirical distribution function $F_{data}(X)$ and the theoretical function $F_0(X)$, given as [1, 2]

$$D = \sup_X \left| F_0(X) - F_{\text{data}}(X) \right|$$

where (for a two-tailed test)

$F_0(X)$ is the cumulative distribution function of the hypothesized distribution

$F_{\text{data}}(X)$ is the empirical distribution function of the observed data

If D is greater than the critical value, the null hypothesis is rejected which means the distributions are not the same [1, 2].

SIDEBAR 2 DURBIN–WATSON TEST

The Durbin–Watson test is a measure of autocorrelation in residuals from regression analysis. Autocorrelation is described as the similarity of a time series over successive time intervals. It can lead to underestimates of the standard error and can lead to believing that the predictors are significant when they are not [1, 2].

The hypothesis for the Durbin–Watson test is [1, 2]:

H_0 : There is no first order autocorrelation of the error terms

H_a : First order autocorrelation of the error terms exists

The assumptions here are that the error terms are normally distributed with a zero mean and the errors are stationary.

The test statistic for Durbin–Watson is calculated as [1, 2]

$$DW = \frac{\sum_{t=2}^{T} (e_t - e_{t-1})^2}{\sum_{t=1}^{T} e_t^2}$$

where e_t are residuals from an ordinary least squares regression.

The Durbin–Watson test reports a score for DW that ranges from 0 to 4 where [1, 2],

$DW = 2$ indicates no autocorrelation

$0 < DW < 2$ indicates a positive autocorrelation

$2 < DW < 4$ indicates a negative autocorrelation

EXERCISE

1. Identify a dataset of your choice to perform Multiple Linear Regression (MLR) analysis. On this dataset perform the following task:
   A. Perform a MLR analyis and report all the performance measures. Discuss about the fit and residuals of the model.
   B. Does your model overfit? If so, what are your options? (**Tips**: Perform Lasso or Ridge or Elastic Net regression)
   Do not forget to define your objectives/research goals, issues in the dataset, any cleaning performed on the dataset (scaling, transformation, removing outliers, imputing missing values, removal of duplicate records, normalization, etc.), document all the steps performed and clearly highlight what inferences you can make or business solutions you can provide based on the analysis strategies and outcomes.

2. If $\beta_0 + x.\beta_1 = -6$ then what is p(y)?
   A. 0
   B. −6
   C. 0.0066928
   D. 0.002472
   E. 0.006737

3. What is the value of $\beta_0 + x.\beta_1$ when p(y) = 0.95?
   A. 4.885324
   B. 2.9444
   C. 0.95
   D. −4.885324
   E. 1
   See the plots below:

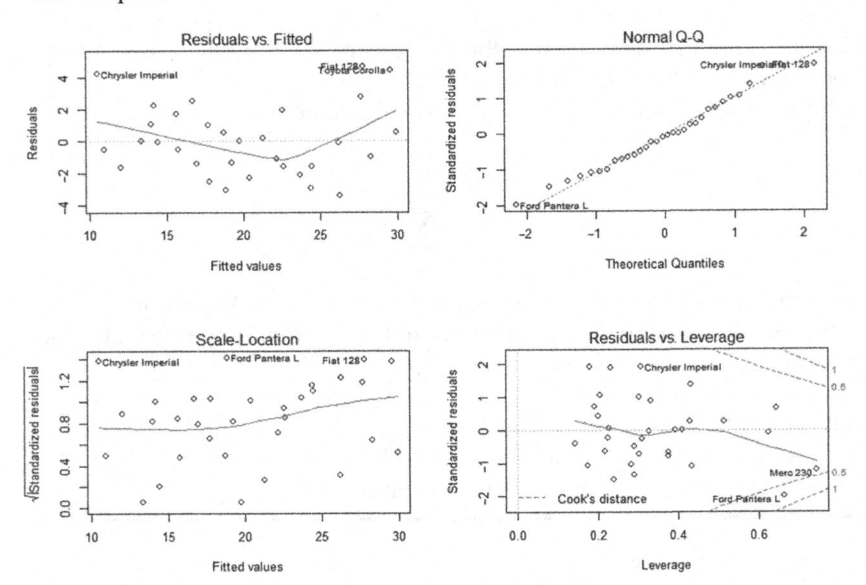

4. Based on the above plots what can you infer about the homoscedasticity of the residuals?
   A. The residuals have equal variance
   B. The residuals have a constant variance
   C. There is no evidence of residuals being homoscedasticity
   D. Both A and B.
   E. All the above.

5. Based on the above plots what can you infer about the distribution of the residuals?
   A. Residuals have bimodal distribution
   B. Residue distribution follows a pareto distribution
   C. Residuals are normally distributed
   D. Residuals are not normally distributed
   E. Residuals are normally distributed but also exhibit a bimodal distribution

```
> vif(mod_1)
      cyl      disp        hp      drat        wt      qsec        vs        am      gear      carb
15.373833 21.620241  9.832037  3.374620 15.164887  7.527958  4.965873  4.648487  5.357452  7.908747
```

6. For the regression model "mod_1" see the VIF's provided above. Based on the VIF values of the predictors which attribute will you first remove from the model?
   A. Cyl
   B. Hp
   C. Draft
   D. disp
   E. Cyl and wt

7. For a regression model with a Durbin–Watson's *d* tests, where $2 < d < 4$ suggests
   A. The model has no autocorrelation
   B. The model suffers from high multicollinearity
   C. The residuals have a constant variance
   D. The model is non-linear
   E. The model has negative autocorrelation

8. In an MLR model the presence of too many predictors can result in
   A. The model exhibiting characteristics for overfitting.
   B. Modeling and accounting for the random noise and the variation in the variable data
   C. Misleading R-squared value
   D. Both A, B and C
   E. Both B and C

9. Download the amusement park dataset (discussed in lesson 4) and extract a random sample containing only 75% of the instances across all the different types of channels. On this dataset determine if customers are more likely to purchase the season pass when it is offered as bundle?

10. One of the techniques listed below is a dimensionality reduction (feature selection) technique
    A. Ridge Regression
    B. Linear Regression
    C. Both Linear and Ridge Regression
    D. Both Ridge and ElasticNet Regression
    E. Lasso Regression

NOTES

1 We disallow the possibility of an "I don't know" or "I don't remember answer" in the questionnaire or survey instrument forcing the respondent to choose either possible response or no answer at all. In the latter case, we would exclude that "data" from the analysis.
2 In statistics white noise refers to a sequence of random variable samples with no mean and limited variance.
3 The square matrix with all ones along the diagonal and zeroes everywhere else.

REFERENCES

1. Chapman, C., Feit, E. M. (2015). *R for Marketing Research and Analytics*, Springer, ISBN 978-3-319-14436-8.
2. Dangeti, P. (2017). *Statistics for Machine Learning*. Packt Publishing Ltd., ISBN 978-1-78829-575-8.
3. Korstanje, J. (2021). "Assumptions of Linear Regression with Implementations in R and Python", retried from https://towardsdatascience.com/assumptions-of-linear-regression-fdb71ebeaa8b, retrieved on April 11, 2022.
4. El-Habil, A. M. (2012). An Application on Multinomial Logistic Regression Model. *Pakistan Journal of Statistics and Operation Research*, VIII(2), 271–291.
5. Hastie, T., Qian, J., Tay, K. (2021). "An Introduction to Glmnet", retrieved from https://glmnet.stanford.edu/articles/glmnet.html, retrieved on April 11, 2022.

5 Unsupervised Learning

Unsupervised learning involves discovering the hidden patterns or structures from a dataset without reference to known or labeled outcomes. These learning methods are often more challenging to implement. Here the outcomes are subjective and there is no simple goal for the analysis. Unsupervised learning is part of exploratory data analysis. It can be hard to assess the results obtained from these methods, since there is no universally accepted way for validating the results. Unsupervised learning methods cannot be applied to regression or classification problem because we have no idea about the values of the output variable which makes it impossible to train the algorithm the way we do with supervised learning. To put in simple terms unsupervised learning methods can be employed to discover the underlying structure of the data [1, 2].

Unsupervised learning methods are becoming popular because, for many applications, a training dataset for many different scenarios may be unavailable. However, even though the purpose of unsupervised learning is to uncover previously unknown patterns in data, the obtained patterns are often very poor approximations of what supervised learning method can achieve. Also, without having the knowledge about the outcome it is hard to determine how accurate these results are and so its applicability to solve real-world problems are not as promising as supervised learning. However, unsupervised methods provide better explainability of patterns compared to supervised techniques. Eventually, patterns obtained through unsupervised learning methods can be used to train supervised learning techniques [1, 2].

Applications of unsupervised learning include [1, 2]:

1. **Clustering**: Allows automatic splitting of the dataset into groups according to similarity.
2. **Anomaly detection**: To identify unusual data points in the dataset such as a fraudulent transaction, discovering faulty piece of hardware/software module, identifying an outlier caused by human error.
3. **Association mining**: Identifying set of items that frequently occur in the dataset such as basket analysis in which the objective is to determine which items are being purchased together for developing effective marketing and merchandising strategies.
4. **Dimensionality reduction**: Commonly used for data preprocessing technique such as reducing the number of features or decomposing the dataset into multiple subsets.
5. **Genomics**: To understand the genomic-wide biological insights from DNA to better understand diseases and people.
6. **Knowledge extraction**: To extract taxonomies of concepts from raw text to generate knowledge graphs.

DOI: 10.1201/9781003278177-5

7. **Segmentation of customers**: In the marketing or banking domain to group similar customers to develop strategies for advertising different products tailored to their needs.

In this chapter, the following topics will be covered to illustrate the potential of unsupervised learning techniques:

- *K*-means clustering.
- Hierarchical clustering.
- Association rule mining.
- K-Nearest Neighbor (KNN).

Later in this chapter, we will show how the K-Nearest Neighbor (KNN) algorithm, which is a clustering algorithm, can be tailored for its application in supervised learning.

K-MEANS CLUSTERING

K-means is an unsupervised classification technique. It is an iterative process of moving the centers of clusters or centroids to the mean position of their constituent points, and reassigning instances to their closest clusters in an iterative manner until there is no significant change in the number of cluster centers possible or the number of iterations reached [1–3].

The Euclidean distance (square-norm) is the cost function for the *k*-means clustering. It is determined by computing the Euclidean distance (see Sidebar 1) between the observations belonging to that cluster with its respective centroid value [1–3].

For example, if there is only one cluster ($k = 1$), then the distances between all the observations are compared with its single mean. Now, if the number of clusters is 2 ($k = 2$), then two means are calculated and a few of the observations are assigned to one cluster (if they are close to this cluster's centroid value) and the remaining observations are assigned to the second cluster based on the proximity computed using the Euclidean distance. Distances are calculated in cost functions by applying the same distance measure, but separately to their cluster centers and is given as [1–3]

$$\sum_{k=1}^{k}\sum_{i\in c_k}\left\|x_i-\mu_k{}^2\right\|$$

K-means is an iterative process of clustering which means that we have to figure out when to stop. There are essentially three stopping criteria that can be adopted. The criteria to stop are if any of the following are satisfied [1–3]:

1. When the centroids of the newly formed clusters do not change.
2. If all the instances (data points) remain in the same cluster.
3. A predetermined number of iterations has reached.

Even after multiple iterations, if the centroids for all the clusters remain the same and the instances remain in the same cluster, we can say that the algorithm is not learning any new pattern and it has converged [1–3].

To illustrate the *k*-means algorithm we will explore segmentation in a customer dataset that can be obtained from this link http://goo.gl/qw303p. The goal of segmentation here is to identify groups of customers that differ from each other in important ways such as with respect to product interest, market participation, and so forth. By understanding the differences among these groups, a strategist can make plans to engage with the groups for effective promotion. The segmentation efforts involve discovering groups in data in order to derive new insights and understand the needs of the particular groups of customers. Therefore, the segmentation can be viewed as a clustering problem [1].

Before we begin clustering let's look at the dataset. The variables in this dataset are the respondent's *age*, *gender*, *household income*, *number of kids*, *home ownership*, *subscription status*, and the assigned *segment memberships*. Now, let's perform clustering on this dataset using R [1].

```
# Let us explore the segmentation data using R
# Read the segmentation data
data_seg <- read.csv("http://goo.gl/qw303p")
# Keep a copy of the raw data
raw_seg <- data_seg
# Now remove the segmentation information
data_seg <- data_seg[,-7]
# Convert all the categorical variables to factor
data_seg$gender <- as.factor(data_seg$gender)
data_seg$ownHome <- as.factor(data_seg$ownHome)
data_seg$subscribe <- as.factor(data_seg$subscribe)
# Generate data summary
summary(data_seg)
```

Upon executing the code, you should see the summary of the dataset similar to Figure 5.1. The segmentation information (*segment*) in this dataset has been removed, since we don't need it for determining the segments of the customers. In fact, the variable *segment* is a response variable that is categorical in nature and is what we are interested in determining. *Segment* has four categories namely "*Moving up*," "*Suburb mix*," "*Travelers*," and "*Urban hip*" (see *geodemographic groups* Sidebar 3). Therefore, here we have a rough idea that there are four different groups of customers, and so we need to cluster the entire dataset into four clusters.

```
      age             gender         income              kids          ownHome       subscribe
 Min.    :19.26   Female:157   Min.    : -5183   Min.    :0.00   ownNo :159   subNo :260
 1st Qu.:33.01   Male  :143   1st Qu.: 39656   1st Qu.:0.00   ownYes:141   subYes: 40
 Median :39.49                 Median : 52014   Median :1.00
 Mean   :41.20                 Mean   : 50937   Mean   :1.27
 3rd Qu.:47.90                 3rd Qu.: 61403   3rd Qu.:2.00
 Max.   :80.49                 Max.   :114278   Max.   :7.00
```

FIGURE 5.1 Summary of the segmentation dataset.

The categorical predictor variables including the *gender*, *ownHome*, and *Subscribe* have been converted to factors for further analysis [1].

Since the *k*-means clustering uses the Euclidean distance as the cost function and works only on numeric data, we need to convert the categorical variables into binary factors, as they can be further coerced to numeric with no alteration of meaning [1].

```
# Now let's duplicate the dataset and convert all the
  categorical (binary) variables to numeric # variables
# First duplicate the dataset
data_seg_num <- data_seg
# use the ifelse to coerced binary to numeric
data_seg_num$gender <- ifelse(data_seg_num$gender=="Male",
  0, 1)
data_seg_num$ownHome <- ifelse(data_seg_num$ownHome=="ownNo",
  0, 1)
data_seg_num$subscribe <- ifelse(data_seg_
  num$subscribe=="subNo", 0, 1)
# show the summary
summary(data_seg_num)
```

Upon executing the code, you should be able to see the summary of the transformed dataset as shown in Figure 5.2.

Now we'll perform the *k*-means clustering to generate 4 ($k = 4$) clusters. When we don't know how many clusters to obtain on a dataset, then methods discussed in Sidebar 2 should be employed to determine the optimal number of clusters.

```
# set seed for repeatability
set.seed(90000)
# Now perform k-means with k=4
seg_k <- kmeans(data_seg_num, centers=4)
# Obtain the summary of the segmented group
seg.summ <- function(data, groups) {
 aggregate(data, list(groups), function(x) mean(as.
  numeric(x)))
 }
# now visualize the groups
seg.summ(data_seg_num, seg_k$cluster)
# Create a boxplot to see how the income of the clustered
  groups vary against each other
boxplot(data_seg_num$income ~ Seg_k$cluster, ylab="Income",
  xlab="Cluster")
```

```
      age            gender            income             kids           ownHome         subscribe
Min.   :19.26   Min.   :0.0000   Min.   : -5183   Min.   :0.00   Min.   :0.00   Min.   :0.0000
1st Qu.:33.01   1st Qu.:0.0000   1st Qu.: 39656   1st Qu.:0.00   1st Qu.:0.00   1st Qu.:0.0000
Median :39.49   Median :1.0000   Median : 52014   Median :1.00   Median :0.00   Median :0.0000
Mean   :41.20   Mean   :0.5233   Mean   : 50937   Mean   :1.27   Mean   :0.47   Mean   :0.1333
3rd Qu.:47.90   3rd Qu.:1.0000   3rd Qu.: 61403   3rd Qu.:2.00   3rd Qu.:1.00   3rd Qu.:0.0000
Max.   :80.49   Max.   :1.0000   Max.   :114278   Max.   :7.00   Max.   :1.00   Max.   :1.0000
```

FIGURE 5.2 Updated summary of the segmentation dataset.

```
  Group.1      age    gender   income     kids  ownHome  subscribe
1       1 42.38909 0.5304348 47799.84 1.434783 0.5304348 0.16521739
2       2 55.40968 0.6000000 89959.96 0.360000 0.8400000 0.12000000
3       3 43.66931 0.5567010 63630.70 1.443299 0.4123711 0.08247423
4       4 29.58704 0.4285714 21631.79 1.063492 0.3015873 0.15873016
```

FIGURE 5.3 Summary of the k-means clustered group with $k = 4$.

Based on the predictors *age* and *income* we can say that the customers in group 4 are relatively young and earn less compared to customers in group 2 who are relatively older and earn more compared to customers in group 1 and group 3. Also, there is a significant difference in the *income* of customers in group 4 when compared to the other groups (see Figure 5.3 and Figure 5.4) [1]. From the boxplot in Figure 5.4 one can clearly infer that there is a substantial difference in income by segment.

Now we'll compare the characteristics of the clustered group against the characteristics of the original segmentation in the dataset to determine how similar they are. In the original dataset we note that the *Travelers* group has a mean age of 57.87, the *Moving up* and the *Suburb mix* groups both have the mean age in the upper 30s and the *Urban hip* group has the mean age of 23.88. Also, note that there is a significant difference in the mean income of the *Travelers* group and the *Urban hip* group, i.e., ~62,000 vs. ~21,000. In addition to that, we can also infer that the *Travelers* group has no kids. Thus, we can say that there is a significant difference between the different groups in the original segmentation dataset (see Figure 5.5). From Figure 5.3, it is highly evident that the cluster 2 is the *Travelers* group and the cluster 4 is the *Urban hip* group. This leaves clusters 3 and 1. Upon matching the characteristics

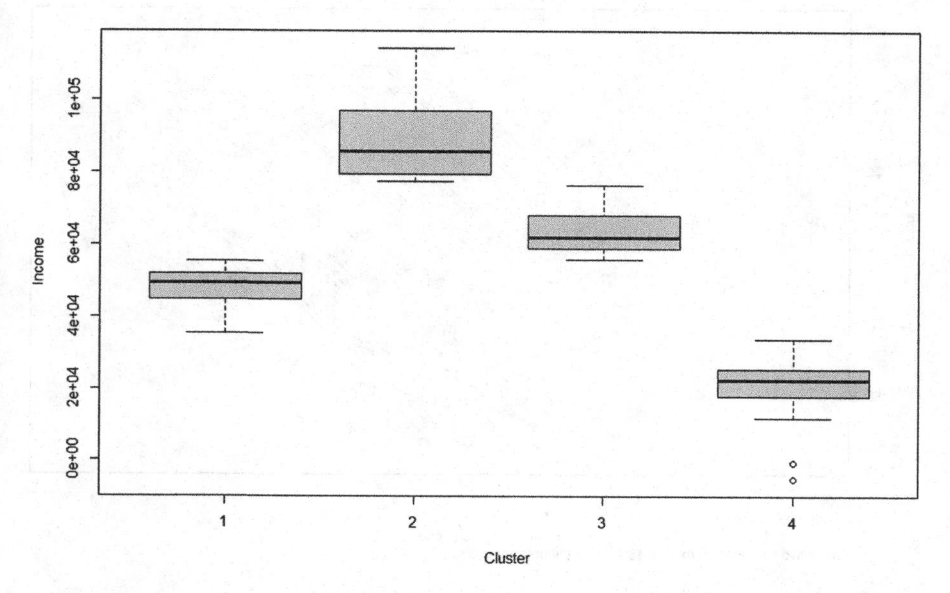

FIGURE 5.4 Boxplot demonstrating the variation in *income* of the clustered group.

	Group.1	age	gender	income	kids	ownHome	subscribe
1	Moving up	36.33114	1.30	53090.97	1.914286	1.328571	1.200
2	Suburb mix	39.92815	1.52	55033.82	1.920000	1.480000	1.060
3	Travelers	57.87088	1.50	62213.94	0.000000	1.750000	1.125
4	Urban hip	23.88459	1.60	21681.93	1.100000	1.200000	1.200

FIGURE 5.5 Characteristics of the segmented groups in the original dataset.

displayed in Figures 5.3 and 5.5, it can be easily guessed that the cluster 3 is the *Suburb mix* group which means that the cluster 1 is the *Moving up* group [1].

Now let's perform a *clusterplot* to determine how much variability in this dataset is being explained by the clustering method and how different are the characteristics of the clustered group. A *clusterplot* is used to perform dimensionality reduction with principal components or multidimensional scaling and plot the observations with cluster membership that has been identified [1].

```
# Construct the clusterplot() after loading the cluster package
library(cluster)
clusplot(data_seg_num, seg_k$cluster, color=TRUE, shade=TRUE,
  labels=4, lines=0, main="K-means cluster plot")
```

From the *clusterplot* (see Figure 5.6), we note that the first two principal components can explain only about 48.5% of the variability in the dataset. Cluster 3 (*Suburb mix*) and cluster 4 (*Urban hip*) are largely overlapping. Cluster 1 (*Moving up*) and 2 (*Travelers*) are modestly differentiated [1].

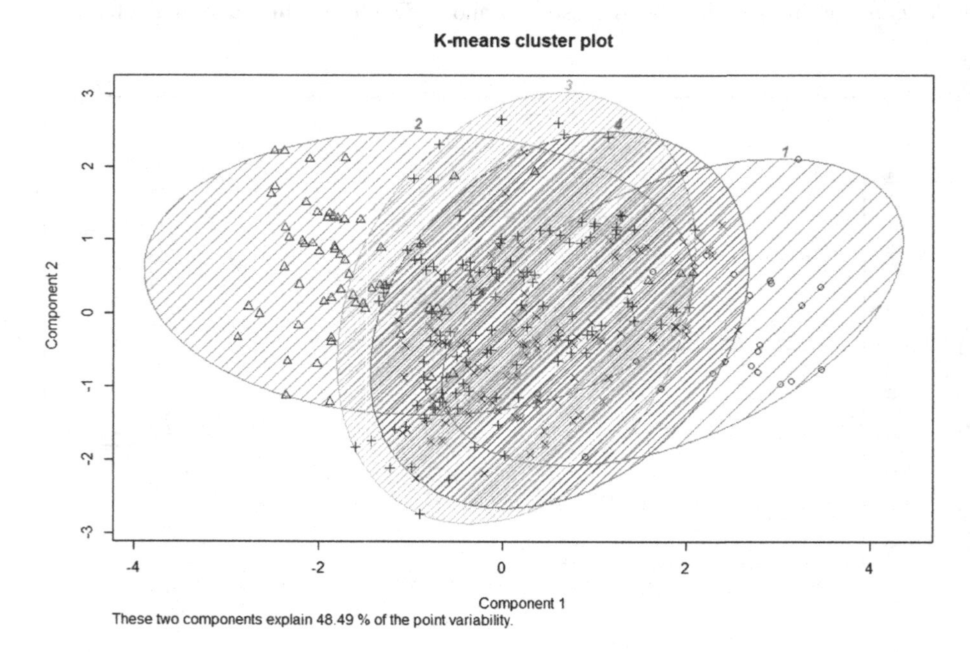

FIGURE 5.6 Clusterplot explaining the characteristics of the clustered groups.

Therefore, the following inferences can be made from this *k*-means clustering exercise [1, 2]:

1. The *Travelers* segment is modestly well differentiated and has the highest average income. That might make it a good target for potential campaign.
2. A limitation of the *k*-means analysis is that it requires specifying the number of clusters which can be very difficult to specify unless the information is known ahead of time.
3. If we were to use *k*-means for the present problem, we would repeat the analysis for $k = 3, 4, 5$, and so forth, and determine which solution gives the most useful result for the business goals. However, there are ways to determine the optimal number of clusters for a given dataset which is further discussed in Sidebar 2.

HIERARCHICAL CLUSTERING

Now let's explore the customer segmentation dataset using another well-known clustering technique called hierarchical clustering. A *hierarchical clustering* algorithm works via grouping data into a tree of clusters. This method begins by treating every data point as a separate cluster. Then the algorithm repeatedly performs the following steps [1]:

1. identify any two clusters that are closer to each other, and
2. merge them in to one cluster.

Both these steps need to be continued until all the clusters have been merged together [1].

Hierarchal clustering or *hclust* is a distance-based algorithm that operates on a *dissimilarity matrix*. The dissimilarity matrix has a $N \times N$ dimension that reports the *distance* between each pair of observations [1]. Here the metric used for determining the distance between the pair of observations is the *Euclidian distance* discussed in Sidebar 1.

In hierarchical clustering the goal is to produce a hierarchical tree of clusters nested together. This tree is also called a *dendrogram*. A dendrogram has the potential to describe the order in which the clusters were merged or spliced. This process of repeatedly joining observations and grouping them is known as an *agglomerative* method. The *agglomerative hierarchical clustering* works as follows [1]:

1. First calculate the similarity of one cluster with all the other clusters. Remember here that each observation (data point) in the dataset is a cluster by itself to begin with.
2. Merge the clusters that are highly similar or close to each other.
3. Recalculate the proximity matrix for each cluster.
4. Repeat steps 2 and 3 until only a single cluster remains.

There is also a *divisive hierarchical clustering* method, which is precisely the opposite of the agglomerative hierarchical clustering. In the divisive hierarchical clustering, all the observations are considered as part of a single cluster and in each iteration, the observations are separated from the cluster which aren't comparable. This is repeated until all the observations taken together become a cluster [1, 2].

Here, we will perform an agglomerative hierarchical clustering on the previously introduced segmentation dataset. A limitation of the hierarchical clustering technique is that the metric Euclidean distance is only defined when the observations are numeric. So, we need to deal with the categorical variables in the segmentation dataset. Remember that there is no way to determine if the predictor is relevant or not for the purpose of clustering. Therefore, we will have to use all the predictors for clustering purposes [1].

In order to perform hierarchical clustering, execute the following R script [1]

```
# Load the cluster library
library(cluster)
# Use the daisy function to rescale the values of all the
  attributes
seg_dist <- daisy(data_seg)
# Now perform hclust() on the dataset using the complete
  linkage method
seg_hc <- hclust(seg_dist, method="complete")
# plot the hierarchical tree or the cluster dendrogram
plot(seg_hc)
```

The *daisy*() function within the cluster package is used to work with mixed data types by rescaling their values. The *hclust*() function from the cluster package can be used in R to perform hierarchical clustering. While performing clustering the *complete* linkage method is adopted which evaluates the distance between every member when combining observations and groups [1]. This is an alternative of the Euclidean distance method. The obtained dendrogram is shown in Figure 5.7.

A dendrogram is interpreted primarily by its height and where the observations are joined. The height represents the dissimilarity between observations that were joined. At the lowest level of the dendrogram tree we see that there are about 2–10 relatively similar observations that are combined, and then those small clusters are successively combined with less similar clusters moving up the tree. It is important here to note that the horizontal ordering of branches is not that important; branches could exchange places with no change in the interpretation [1].

Let's cut the dendrogram tree to clearly visualize how observations have been combined to form clusters using the following R code [1]:

```
# Let us take a closer look - Cut the tree at the height of 0.5
plot(cut(as.dendrogram(seg_hc), h=0.5)$lower[[1]])
# Similar instances - Instance 101 and 107 looks very similar
data_seg[c(101, 107), ]
```

Upon executing the code, we obtain the tree shown in Figure 5.8. Here we can see closer detail of the clustering, for example, we can see that customer 101 and 107 have been clustered together [1].

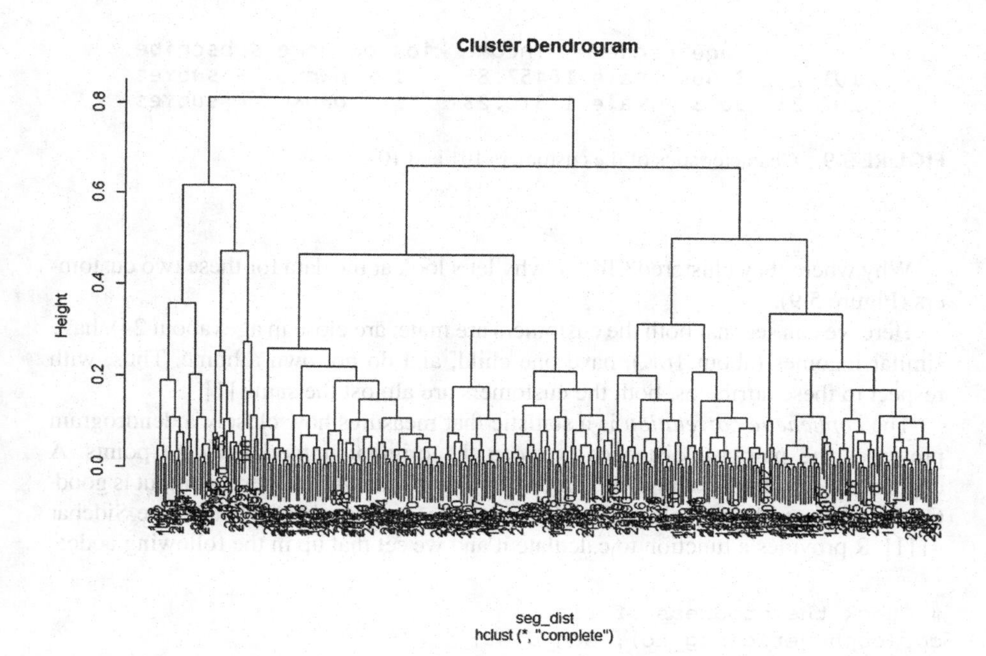

FIGURE 5.7 Dendrogram obtained on the segmentation dataset.

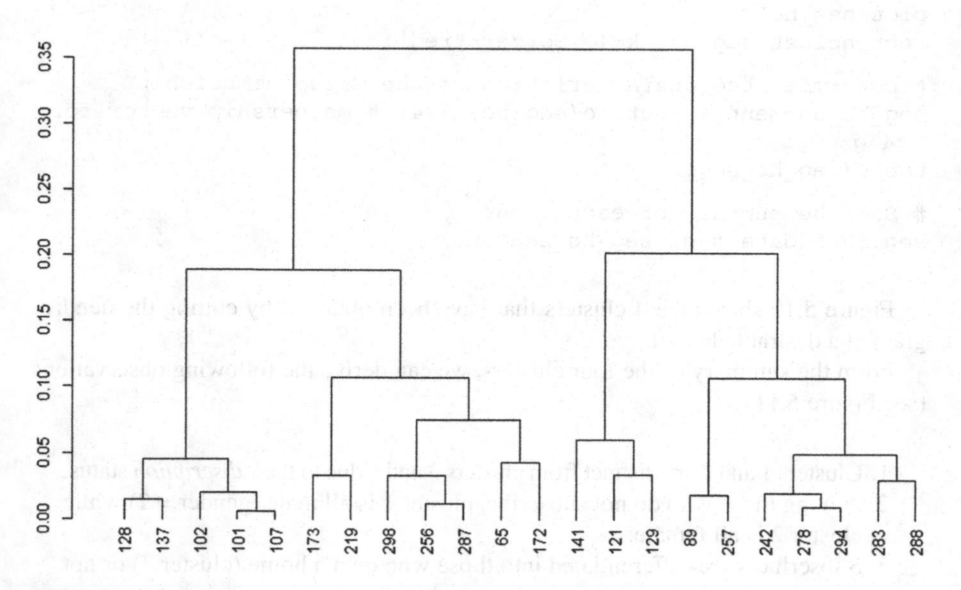

FIGURE 5.8 Dendrogram tree cut at height of 0.5.

```
         age gender   income kids ownHome subscribe
101 24.73796   Male 18457.85    1   ownNo    subYes
107 23.19013   Male 17510.28    1   ownNo    subYes
```

FIGURE 5.9 Characteristics of the customers 101 and 107.

Why where they clustered? To see why let's look at the data for these two customers (Figure 5.9).

Here we can see that both the customers are male, are close in age (about 24), have similar incomes (about 18K), have one child, and do not own a home. Thus, with respect to these attributes, both the customers are almost the same [1].

The *cophenetic correlation* is a statistic that measures how closely a dendrogram preserves the pairwise distances between the original unmodeled data points. A cophenetic correlation of higher than 0.768 suggests that the clustering output is good. Cophenetic correlation value of models close to 1 are considered better (see Sidebar 4) [1]. R provides a function to calculate it and we set that up in the following code:

```
# Check the goodness of fit
cor(cophenetic(seg_hc), seg_dist)

# Infer the groups in the segmentation data
# A dendrogram can be cut into clusters at any height desired,
  resulting in different groups.
plot(seg_hc)
rect.hclust(seg_hc, k=4, border="red")

# Determine the characteristics of the Group membership
seg_hc_segment <- cutree(seg_hc, k=4) # membership vector for
  4 groups
table(seg_hc_segment)

# See the summary of each group
seg.summ(data_seg, seg_hc_segment)
```

Figure 5.10 shows the 4 clusters that have been obtained by cutting the dendrogram at a desirable height.

From the summary of the four clusters, we can derive the following observations (see Figure 5.11)

1. Clusters 1 and 2 are distinct from clusters 3 and 4 due to the *subscription* status.
2. Among those who do not subscribe, cluster 1 is all male (gender = 2) while cluster 2 is all female.
3. Subscribers are differentiated into those who own a home (cluster 3) or not (cluster 4).

Next, we will discuss about another well-known unsupervised learning technique.

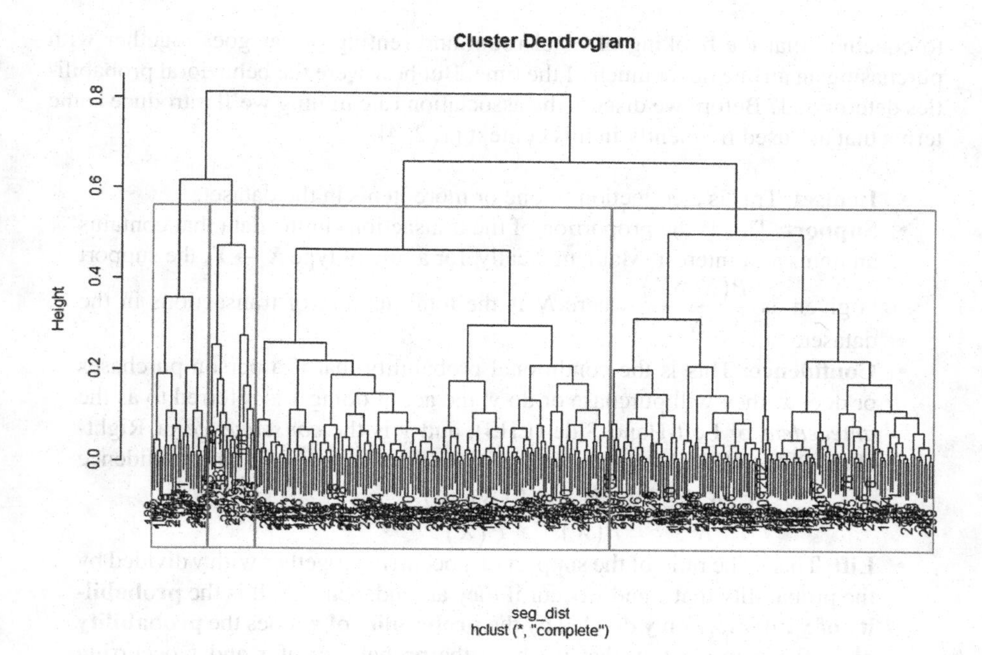

FIGURE 5.10 Dendrogram tree cut at a height in order to create 4 clusters.

	Group.1	age	gender	income	kids	ownHome	subscribe
1	1	40.78456	2.000000	49454.08	1.314516	1.467742	1
2	2	42.03492	1.000000	53759.62	1.235294	1.477941	1
3	3	44.31194	1.388889	52628.42	1.388889	2.000000	2
4	4	35.82935	1.545455	40456.14	1.136364	1.000000	2

FIGURE 5.11 Characteristics of the segmented groups as determined by hierarchical clustering [1].

ASSOCIATION RULE MINING

Association rule mining is a data mining technique that has the purpose of finding the optimal combination of products or services that allows service providers to exploit this knowledge to provide recommendations, optimize product placement, or develop programs that take advantage of cross-selling. In short, the idea is to identify which products pair well. Association rule making is among the first "classical" data mining techniques introduced in the early 1990s. Many tools to implement these techniques were soon introduced, revolutionizing the marketing techniques used by product manufacturers, stores, and service providers [1, 2, 4].

Consider the following example, suppose an analysis of traveler behavior determines that if a customer buys an airplane ticket, then there is a 50% probability that they will also book a hotel room, and if they go on to book a hotel room, then there is a 30% probability that they will rent a car. From these statistics it would be natural to conclude that the booking of hotel room and renting of car goes together with purchasing an airline ticket much of the time. But how were the behavioral probabilities determined? Before we discuss the association rule mining we'll introduce some terms that are used frequently in this context [1, 2, 4].

- **Itemset**: This is a collection of one or more items in the dataset.
- **Support**: This is the proportion of the transactions in the data that contains an itemset of interest. Mathematically, for a rule of type $X \rightarrow Y$, the support is given as $\frac{P(X \cap Y)}{N}$ where N is the total number of transactions in the dataset.
- **Confidence**: This is the conditional probability that if a person purchases or does x, they will purchase or do y; the act of doing x is referred to as the *antecedent* or Left-Hand Side (LHS), and y is the *consequence* or Right-Hand Side (RHS). Mathematically, for a rule of type $X \rightarrow Y$, the confidence is given as $\frac{P(X \cap Y)}{N} \times \frac{N}{P(X)} = \frac{P(X \cap Y)}{P(X)}$
- **Lift**: This is the ratio of the support of x occurring together with y divided by the probability that x and y occur if they are independent. It is the **probability of x intersection y** divided by the **probability of x** times the **probability of y**; for example, say that we have the probability of x and y occurring together as 10% and the probability of x is 20% and y is 30%, then the lift would be 1.67. Mathematically, for a rule of type $X \rightarrow Y$, the lift is given as $= \frac{P(X \cap Y)}{P(X) \times P(Y)}$

Let's consider an example to illustrate these concepts. In Figure 5.12 we see 5 shopping carts each containing 3 out of 5 items, namely $A, B, C, D,$ and E [4].

Let us consider few rules and determine the support, confidence, and lift of the respective rules (see Table 5.1) [4].

Now let's discuss the *apriori* algorithm, which we will be using to perform association rule mining on the grocery cart dataset. With *apriori*, the principle is that, if an *itemset* is frequent, then all its subsets must also be frequent. A minimum frequency (support) is determined prior to executing the algorithm, and once established, the algorithm will run as described in Table 5.2 [1, 2, 4].

FIGURE 5.12 5 shopping carts each containing 3 items. (Source: Figure adapted from Li, S. (2017). "A gentle introduction on Market Basket Analysis - Association Rules," retrieved from https://towardsdatascience.com/a-gentle-introduction-on-market-basket-analysis-association-rules-fa4b986a40ce, retrieved on September 15, 2022.)

TABLE 5.1
Support, Confidence, and Lift of the Given Rules

Rule	Support	Confidence	Lift
$A \to D$	$\frac{P(A \cap D)}{N} = \frac{2}{5}$	$\frac{P(A \cap D)}{P(A)} = \frac{2}{3}$	$\frac{P(A \cap D)}{P(A) \times P(D)} = \frac{10}{9}$
$C \to A$	$\frac{P(C \cap A)}{N} = \frac{2}{5}$	$\frac{P(C \cap A)}{P(C)} = \frac{2}{4}$	$\frac{P(C \cap A)}{P(C) \times P(A)} = \frac{5}{6}$
$A \to C$	$\frac{P(A \cap C)}{N} = \frac{2}{5}$	$\frac{P(A \cap C)}{P(A)} = \frac{2}{3}$	$\frac{P(A \cap C)}{P(A) \times P(C)} = \frac{5}{6}$
$B,C \to D$	$\frac{P(B,C \cap D)}{N} = \frac{1}{5}$	$\frac{P(B,C \cap D)}{P(B,C)} = \frac{1}{3}$	$\frac{P(B.C \cap D)}{P(B,C) \times P(D)} = \frac{5}{9}$

TABLE 5.2
Apriori Algorithm

Let $k = 1$ (the number of items)
Generate itemset of a length that are equal to or greater than the specified support
Iterate $k + (1 \ldots n)$, pruning those that are infrequent (less than the support)
Stop the iteration when no new frequent itemset are identified

Once the ordered summary of most frequent *itemsets* has been obtained, the analysis process can continue by examining the confidence and lift in order to identify the associations of interest [1, 2, 4].

Here we will explore the grocery dataset using the *apriori* algorithm. This dataset consists of transactions over a 30-day period from a real-world grocery store and consists of 9,835 different purchases with items aggregated to 169 categories [1].

In R, the *arules* package has an implementation of the *apriori* algorithm. The *arules* package can be used for mining association rules and discovering frequent itemset. Before performing the association rule mining, though, the packages namely *arules* and *arulesViz* should be installed. We use the following R code [1]:

```
# Install the required packages

install.packages("arules")
install.packages("arulesViz")
# Load the Groceries dataset
library(arules)
library(arulesViz)
data(Groceries)
# Get the first 5 transactions from the dataset
inspect(Groceries[1:5])
```

After running the code the first 5 transactions from the groceries dataset are shown in Figure 5.13.

```
      items
  [1] {citrus fruit,
       semi-finished bread,
       margarine,
       ready soups}
  [2] {tropical fruit,
       yogurt,
       coffee}
  [3] {whole milk}
  [4] {pip fruit,
       yogurt,
       cream cheese ,
       meat spreads}
  [5] {other vegetables,
       whole milk,
       condensed milk,
       long life bakery product}
  >
```

FIGURE 5.13 Transactions of items in the groceries dataset.

Now let's execute the apriori algorithm with a support of 0.001 and a confidence of 0.9 using the following R code [1].

```
# Now model the dataset
rules <- apriori(Groceries, parameter = list(supp = 0.001,
  conf = 0.9, maxlen=4))
#How many rules do I have?
rules
# show values up to 2 decimal places
options(digits = 2)
# Sort rules by lift
rules <- sort(rules, by = "lift", decreasing = TRUE)
inspect(rules[1:5])
```

Upon execution we obtain a total of 67 rules have with a max length of 4, i.e., the total number of items in a rule is 4. Figure 5.14 lists the first 5 rules after sorting all the 67 rules by the decreasing order of the lift [1].

We can see that customers who buy liquor and red/blush wine also buy bottled beer (see the first rule in Figure 5.13). This rule has a high lift (11.23) and also high confidence (90.4%). A lift greater than 1 means that there is a strong association between X (LHS) and Y (RHS) for the given rule $X \rightarrow Y$. All the rules in Figure 5.13 have a strong association between the LHS and RHS [1].

Next, we will focus our discussion on the KNN algorithm and its application.

```
    lhs                                            rhs                support confidence coverage lift count
[1] {liquor, red/blush wine}                    => {bottled beer}     0.0019  0.90       0.0021   11.2 19
[2] {root vegetables, butter, cream cheese }    => {yogurt}           0.0010  0.91       0.0011    6.5 10
[3] {citrus fruit, root vegetables, soft cheese} => {other vegetables} 0.0010  1.00       0.0010    5.2 10
[4] {pip fruit, whipped/sour cream, brown bread} => {other vegetables} 0.0011  1.00       0.0011    5.2 11
[5] {butter, whipped/sour cream, soda}          => {other vegetables} 0.0013  0.93       0.0014    4.8 13
```

FIGURE 5.14 Top 5 Apriori rules sorted by the decreasing order of lift.

K-NEAREST NEIGHBORS

The *K-nearest neighbors* (*KNN*) is considered both an unsupervised and a supervised learning algorithm. KNN can be used for performing both regression and classification. The KNN tries to predict the class (classification) for the test instances by calculating the distance between the test instance and all the training instances. Once the distance is computed it selects the *K* number of points which is close to the test instance [2, 5–7].

The KNN algorithm calculates the probability of the test instance belonging to the classes of "*K*" training instances and the class that holds the highest probability will be selected. In the case of regression, the value is the mean of the "*K*" selected training points [2, 5–7].

Implementation of KNN can be explained using the following steps [2, 5–7]:

1. Select the number of neighbors, i.e., the number k.
2. Calculate the Euclidean distance of k number of neighbors from the test instance that needs to be classified.
3. Take the k nearest neighbors as per the calculated Euclidean distance.
4. Among these k neighbors, count the number of the data instances in each category.
5. Assign the new data instance to that category for which the number of the neighbors is maximum.

The performance of the KNN algorithm depends on the chosen value of k. A small value of k should be avoided when using the KNN for classification purpose. One way of determining an optimal value of k is considering $k = \sqrt{N}$ where N is the total number of instances in the sample [2, 5–7].

Let's consider a simple example to illustrate the KNN algorithm. A tissue paper is classified either as *good* or *bad* based on two attributes namely the *acid durability* and *strength* as shown in Table 5.3 [6].

In this dataset we would like to determine the classification (class) of the tissue paper with the *acid durability* = 3 and *strength* = 7. Here we will consider $K = 3$ [6].

Using the Euclidean distance, we compute the distance of the test data point (*acid durability* = 3 and *strength* = 7) from the other data points described in Table 5.3 (see Table 5.4) [6].

Now let's determine the 3-nearest neighbors of the test data point (see Table 5.5).

TABLE 5.3
Example Dataset to Demonstrate the KNN Algorithm

Acid Durability	Strength	Class
7	7	Bad
7	4	Bad
3	4	Good
1	4	Good
3	7	?

TABLE 5.4
Euclidean Distance of Data Points from the Test Data Point

Acid Durability	Strength	Class	Distance
7	7	Bad	$\sqrt{(7-3)^2+(7-7)^2}=4$
7	4	Bad	$\sqrt{(7-3)^2+(4-7)^2}=3$
3	4	Good	$\sqrt{(3-3)^2+(4-7)^2}=3$
1	4	Good	$\sqrt{(1-3)^2+(4-7)^2}=3.60$
3	7	?	

TABLE 5.5
3-Nearest Neighbors of the Test Data Point

Acid Durability	Strength	Class	Distance	3-Nearest Neighbor
7	7	Bad	$\sqrt{(7-3)^2+(7-7)^2}=4$	NO
7	4	Bad	$\sqrt{(7-3)^2+(4-7)^2}=3$	YES
3	4	Good	$\sqrt{(3-3)^2+(4-7)^2}=3$	YES
1	4	Good	$\sqrt{(1-3)^2+(4-7)^2}=3.60$	YES
3	7	?		

From Table 5.5 it can be inferred that the test data point is closest to data points (*acid durability* = 7 and *strength* = 4), (*acid durability* = 3 and *strength* = 4), and (*acid durability* = 1 and *strength* = 4) which are classified as bad, good, and good, respectively, since the majority of the closest neighbors of the test data points are classified as good. The tissue paper with *acid durability* = 3 and *strength* = 7 (the test data point) can be classified as good quality paper [6].

Now that we have a good understanding of the KNN algorithm let's classify the instances of the Iris dataset using the KNN algorithm.

The steps listed below (including the R script) can be followed to perform the KNN clustering. In order to perform KNN we will use the implementation of KNN in the *class* library of R. To begin, obtain the Iris dataset. Iris is the best-known dataset to be found in the pattern recognition literature. This dataset contains 3 classes of 50 instances each for types of plants namely the *Iris Setosa*, the *Iris Versicolour*, and the *Iris Virginica*. The predictors in this dataset are the *sepal length*, *sepal width*, *petal length*, and *petal width*. In total, this dataset has 150 instances [2, 7].

```
              model1
iris.test.target setosa versicolor virginica
      setosa          0          0         0
      versicolor      0          0         0
      virginica       0          2        18
```

FIGURE 5.15 Classification output of the KNN algorithm on the Iris test dataset.

```
# Import the library class
library(class)
# Obtain the Iris dataset
data(iris)

# normalize the dataset
normalize <- function(x){
 return ((x-min(x))/(max(x)-min(x)))
}
iris.new<- as.data.frame(lapply(iris[,c(1,2,3,4)],normalize))
head(iris.new)

# subset the dataset
iris.train<- iris.new[1:130,]
iris.train.target<- iris[1:130,5]
iris.test<- iris.new[131:150,]
iris.test.target<- iris[131:150,5]
```

The next step is to normalize or scale all the four predictors in the Iris dataset. After normalization the dataset is split in to train and test dataset. The first 130 instances in the dataset are used to train the KNN classifier and the remaining 20 instances are considered as test instances which will be classified using the KNN classifier. All the 20 instances belong to the *Iris Virginica* class. With this knowledge it will be interesting to see how many instances are correctly classified by the KNN algorithm [2, 7]. To perform KNN we will have to construct a model called *model1* as shown in the following R code [2, 7]:

```
model1<- knn(train=iris.train, test=iris.test, cl=iris.train.
  target, k=10)
# Now let us see how many instances are correctly predicted
table(iris.test.target, model1)
```

From Figure 5.15 it can be inferred that 18 out of 20 instances were correctly predicted as *Iris Virginica.*

SUMMARY

The focus of this chapter was to introduce the readers to the concept of unsupervised learning. This chapter introduces different unsupervised learning techniques including the *k*-means clustering, hierarchical clustering, association rule mining, and the *k* Nearest Neighbors. Through simulated and real-world examples, it has been demonstrated how these machine learning techniques can discover the hidden patterns or structures from a dataset without any reference. However, it should be remembered that the outcomes are subjective and there is no way to validate the results. In the next chapter we will discuss about the supervised learning techniques.

SIDEBAR 1 CLUSTERING DISTANCE MEASURES

To classify the observations into groups there is a need for computing the distance or the (dis)similarity between each pair of observations. There are many methods to calculate this distance information. Therefore, the choice of distance measures is a critical step in clustering.

The classical methods for measuring distances are *Euclidean* and *Manhattan distances*, which are defined as follow [2]:

Euclidean distance:

$$d_{euc}(x,y) = \sqrt{\sum_{i=1}^{n}(x_i - y_i)^2}$$

Manhattan distance:

$$d_{man}(x,y) = \sum_{i=1}^{n}\left|(x_i - y_i)\right|$$

SIDEBAR 2 ELBOW CURVE METHOD AND SILHOUETTE ANALYSIS

To determine the optimal number of clusters the following techniques are be exploited [1, 2].

Elbow Curve Method—First perform *K*-means clustering with different values of K. Second, for each of the K values, one should calculate the average distance of all data points to the centroid. Finally, plot all these points and find the point where the average distance from the centroid falls suddenly.

Silhouette Analysis—This is a measure of how similar a data point is within-cluster (cohesion) compared to other clusters (separation). The equation for calculating the silhouette coefficient for a particular data point is given as

$$S(i) = \frac{b(i) - a(i)}{\max\{a(i), b(i)\}}$$

where,

$S(i)$ is the silhouette coefficient of the data point i.

$a(i)$ is the average distance between i and all the other data points in the cluster to which i belongs.

$b(i)$ is the average distance from i to all clusters to which i does not belong.

The average silhouette is then calculated as $mean(S(i))$.

SIDEBAR 3 GEODEMOGRAPHIC SEGMENTATION

Geodemographic Segmentation refers to certain techniques for characterizing neighborhoods based on the assumption that people who live close by have similar demographic, socioeconomic, and lifestyle traits. Census data along with other public record information such as property values and taxes, voting results, and court records are also used. K-means and other clustering algorithms (such as fuzzy clustering) are typically used to create the segments The created segments are often given colorful and evocative names. For example, from the previously discussed market clustering case, *"Moving up,"* might describe upwardly mobile younger people, *"Suburb mix,"* a heterogeneous community of various ages and professions, *"Travelers,"* persons who tend not to stay in the same place for long and *"Urban hip,"* younger city dwellers. Obviously these groups all have very different needs and desires, spending ability, and so on.

Geodemographic segmentation revolutionized marketing when it was introduced in the early 1980s. Aside from marketing geodemographic segmentation is used for political campaigning, public health planning and tracing, policing, and more.

To see geodemographic segmentation up close try typing a zip code into the following mapping tool https://claritas360.claritas.com/mybestsegments/#zipLookup. Try your own zip code and consider if this tool is accurate in your case. There are many such commercial implementations of geodemographic segmentation used by marketing professionals, urban planners, public health officials, and more, all over the world.

SIDEBAR 4 COPHENETIC CORRELATION COEFFICIENT

It is a measure of how accurately and reliably a dendrogram preserves the pairwise distance between the original unmodeled data points. Let us suppose that the original data points $\{X_i\}$ have been modeled using a clustering method to produce a dendrogram $\{T_i\}$, i.e., a simplified model in which data that are close have been grouped into a hierarchical tree. Now let us define the following distance measures [1, 2, 8].

The Euclidean distance between the *i*th *and j*th observations given as $x(i,j) = |x_i - x_j|$

The dendrogrammatic distance between the model points T_i and T_j, i.e., the distance or the height of the node at which these two points are first joined together given as $t(i,j)$

If $\bar{x}$ is the average of the $x(i,j)$ and if $\bar{t}$ is the average of the $t(i,j)$ then the cophenetic correlation coefficient c is given as

$$c = \frac{\sum_{i<j}\left[x(i,j)-\bar{x}\right]\left[t(i,j)-\bar{t}\right]}{\sqrt{\sum_{i<j}\left[x(i,j)-\bar{x}\right]^2 \sum_{i<j}\left[t(i,j)-\bar{t}\right]^2}}$$

If $c = 1$ then it implies that the dendrogram accurately preserves the pairwise distance between the original data points [1, 2, 8].

EXERCISE

1. Download the segmentation dataset (discussed in lesson 5) and extract a random sample containing 80% of the instances from this dataset. First, summarize this dataset. Second, on this dataset use any ONE unsupervised (clustering) technique and discuss about the outcome of the analysis. Clearly define your objectives/research goals, issues in the dataset, any cleaning performed on the dataset (scaling, transformation, removing outliers, imputing missing values, removal of duplicate records, normalization, etc.), document all the steps performed, and clearly highlight what inferences you can make or business solutions can be provided based on your strategies and outcomes.

2. Cluster the following five points (with (x, y) representing locations in a 2D plane) into two clusters:
   A1(5, 8), A2(7, 5), A3(6, 4), A4(1,2), A5(4, 9)
   Note: Use the Euclidian distance as the distance measure

3. Using the *k*-means clustering algorithm perform multivariate outlier detection on the 50% of the instances randomly chosen from the Iris dataset. Clearly list all the identified outliers. Provide the R/Python script used for performing the multivariate outlier detection and clearly state all your assumptions.

4. What is the difference between the Euclidean distance and the Manhattan distance measurement? By using an example clearly show when you would use one type of distance measurement technique versus another. For calculation purpose you can use the calculator. Consider a small dataset (5 data points) for illustration purposes.

5. Consider the Market basket transactions provided below.

Transaction ID	Items Bought
1	{Milk, Beer, Diapers}
2	{Bread, Butter, Milk}
3	{Milk, Diapers, Cookies}
4	{Bread, Butter, Cookies}

Transaction ID	Items Bought
5	{Beer, Cookies, Diapers}
6	{Milk, Diapers, Bread, Butter}
7	{Bread, Butter, Diapers}
8	{Beer, Diapers}
9	{Milk, Diapers, Bread, Butter}
10	{Beer, Cookies}

Using the Apriori algorithm provide your solutions for the following

A. What is the maximum number of size-3 itemset that can be derived from this data set?

B. Find an itemset (of size 2 or larger) that has the largest support

C. Find a pair of items, a and b, such that the rules $\{a\} \dashrightarrow \{b\}$ and $\{b\} \dashrightarrow \{a\}$ have the same confidence

D. Find an itemset (of size 2 or larger) that has the largest lift

6. Using the hierarchical clustering algorithm perform multivariate outlier detection on the **mtcars** dataset. Clearly list all the identified (car types) outliers. Provide the R/Python script used for performing the multivariate outlier detection and clearly state all your assumptions.

7. Using the KNN algorithm perform multivariate outlier detection on the **mtcars** dataset and compare the results with the outliers detected using the hierarchical clustering algorithm in question 6. Do you see any difference in the output of the two algorithms?

8. You have built an algorithm that is used to separate data into two groups. During the analysis, you have discovered that, actually, data needs to be separated in to four groups. Which of the following algorithms can be used without modification to support for the change in the specification of the problem? (Only one of the following options is correct)
   A. Associative classification
   B. Hierarchical cluster analysis
   C. Sequence analysis
   D. K-means cluster analysis
   E. K-medoids cluster analysis

9. Which of the following is true about agglomerative hierarchical clustering algorithms?
   A. They can be applied only to numerical data
   B. Always merge the pair of clusters that are the closest to each other
   C. They will revisit data points which have been assigned to clusters
   D. They are a special case of the k-means clustering algorithms

10. Which of the following is true about the *k*-means algorithm? Please choose all the options that apply.
    A. Is typically done in Excel or similar software
    B. Can converge to different final clusterings, depending on initial choice of representatives
    C. It is difficult to implement due to multiple special cases
    D. It always converges to a clustering that minimizes the mean-square vector-representative distance

REFERENCES

1. Chapman, C., Feit, E. M. (2015). *R for Marketing Research and Analytics*, Springer, ISBN 978-3-319-14436-8.
2. Dangeti, P. (2017). *Statistics for Machine Learning*. Packt Publishing Ltd., ISBN 978-1-78829-575-8.
3. Boehmke, B. (n.d.). "K-means Clustering Analysis", retrieved from https://uc-r.github.io/kmeans_clustering, retrieved on May 1, 2022.
4. Li, S. (2017). "A Gentle Introduction on Market Basket Analysis- Association Rules", retrieved from https://towardsdatascience.com/a-gentle-introduction-on-market-basket-analysis-association-rules-fa4b986a40ce, retrieved on September 15, 2022.
5. Christopher, A. (2021). "K-Nearest Neighbor", retrieved from https://medium.com/swlh/k-nearest-neighbor-ca2593d7a3c4, retrieved on May 1, 2022.
6. Teknomo, K. (n.d.). "K-Nearest Neighbors Tutorial", retrieved from https://people.revoledu.com/kardi/tutorial/KNN/KNN_Numerical-example.html, retrieved on May 1, 2022.
7. Nitika. (2017). "KNN Classification Demo", retrieved from https://rpubs.com/Nitika/kNN_Iris, retrieved on May 1, 2022.
8. "Cophenetic correlation." Wikipedia, Wikimedia Foundation, 10 September 2021, retrieved from https://en.wikipedia.org/wiki/Cophenetic_correlation.

6 Supervised Learning

The previous chapter discussed unsupervised learning. This chapter is a formal introduction to supervised learning. *Supervised learning* or *supervised machine learning* is a subcategory of machine learning (ML) and artificial intelligence (AI). In supervised learning a labeled dataset is used to train the algorithms to classify and to predict the outcomes of unlabeled instances. Based on the input data the parameters of the model get adjusted until the fit is appropriate, which occurs as part of the cross-validation process.

Supervised learning helps organization to solve a variety of real-world problems that require classification. In this chapter, we will discuss four important classification algorithms, namely, Artificial Neural Networks, Random Forest, AdaBoost, and eXtreme Gradient Boosting.

INTRODUCTION TO ARTIFICIAL NEURAL NETWORKS

The *Artificial Neural Network* (*ANN*) classifier models the relationship between a set of input and output signals using an interconnected network of artificial neurons (or nodes). The structure and functionality of the artificial neurons were discussed in Chapter 2. In short, each neuron of the ANN has a set of inputs, each of which is given a specific weight. These neurons compute a function on the weighted inputs. Based upon the function type (sigmoid, tanh, ReLU, etc.) the neurons determine an output signal which is the function of a combination of the weighted inputs [1, 2].

A neuron with n inputs, i.e., $x_1, x_2, x_3, \ldots, x_n$ with the weights for each input given as $w_1, w_2, w_3, \ldots, w_n$, contributes a greater or lesser amount to the sum of input signals given as $\sum_{i=1}^{n} w_i x_i$. Therefore, the output signal $y(x)$ when passed through the activation function f is given as $y(x) = f\left(\sum_{i=1}^{n} w_i x_i\right)$ [1, 2]. For a further discussion of the history of artificial neural network, the precursor to all neural networks, see Sidebar 1.

Now let's discuss the parameters required for building the ANN model. These parameters are [1–3]:

Activation function: This function plays a major role in aggregating the input signals into an output signal that is propagated to the other neurons in the network. Activation functions were discussed in detail in Chapter 2.

Network architecture: The network architecture deals with the number of layers and the number of neurons in each layer. The greater the number of layers and neurons in the network, the larger the non-linear decision boundary. On the other hand, the fewer the number of layers and neurons, the less flexible but more robust is the model.

Optimization algorithm for training: The optimization algorithm plays a critical role in determining how quickly and accurately the convergence will take place to the best optimal solutions.

DOI: 10.1201/9781003278177-6

Here, we'll discuss about the forward and back propagation method of calculating the gradient of neural network parameters for a two hidden layer neural networks, i.e., a Multi-Layer Perceptron (MLP) thus laying the foundation for deep neural networks (DNN) that will be discussed in detail in Chapter 8.

FORWARD AND BACKWARD PROPAGATION METHODS

In an ANN the number of neurons in the input layer is based on the number of independent or input variables, whereas the number of neurons in the output layer depends on the number of classes that the model needs to predict. In the MLP, we assume that there are three neurons in both the hidden layers. The weights and biases of the MLP are initialized with random numbers so that in both the forward and backward passes these weights and biases can be updated in order to minimize the total error. Now let's look at forward and backward propagation methods [1, 2].

During forward propagation, the features that are input to the network are fed through the layers to produce the output activation. In the first hidden layer, the activation is obtained as the combination of the bias weights and the weighted combination of the input values. If the activation exceeds the threshold, it will trigger to the next layer, otherwise the output signal will be a zero input to the next layer. The bias is used here to control the triggering points. Whenever the weighted combination of the signal is low, the bias will compensate by adjusting the aggregated value thus serving as a trigger for the next level. Once the neurons are determined in the first hidden layer, then the neurons in the next layer will need to be determined in a similar fashion, i.e., using the activation output of the hidden neurons from the first layer plus the bias [1, 2].

In the output layer the outputs are calculated in the same way, i.e., by using the outputs obtained from the second hidden layer (taking the weighted combination of weights). Once the output from the model is obtained, a comparison is performed with the actual value to determine the errors. This error is then backpropagated across the network in order to correct the weights of the entire neural network [1, 2].

To start, the derivative of the output value is taken and multiplied by the error component that was obtained from differencing the actual value with the model output: In a similar manner the error is backpropagated from the second hidden layer as well. The errors are then computed for the neurons in the second hidden layer [1, 2].

Finally, the errors are calculated for the neurons in the first hidden layer based on the errors obtained from the neurons in the second hidden layer. Once all the neurons in the first hidden layer are updated, weights between the inputs and the first hidden layer also need to be updated. In a similar way, all the weights get updated until the convergence takes place, or the number of iterations specified is reached [1, 2].

Now that we have described how the forward and backward propagation methods work to stabilize the weights and bias of the network, we can explore the different architectures of ANN.

ARCHITECTURAL TYPES IN ANN

One of the most important aspects of the ANN architecture is the interconnections. Interconnections are the arrangements of the processing elements in ANN. In all network architectures the input and the output layers are common. The third layer, also referred to as the *hidden layer*, is the distinguishing feature between different architectures. The hidden layer acts as a black box to the users as the neurons in these layers are hidden from them. Increasing the number of hidden layers boosts the system's computational and processing power along with the system's complexity [1, 2]. You will learn more about it in Chapters 8–10 where the deep learning–based classifiers will be discussed in detail.

There are five basic types of neuron interconnection architectures, namely, the single-layer feed-forward network, the multilayer feed-forward network, the single node with its own feedback, the single-layer recurrent network, and the multilayer recurrent network. Let's briefly discuss each of these [1–3].

In the single-layer feed-forward network there are only two layers, i.e., the input layer and the output layer. In the neurons (nodes) of the input layer, different weights are applied and cumulative effect per node is taken into consideration to form the output layer. The neurons in the output layer collectively compute the output signal [1–3] (see Figure 6.1).

In the multilayer feed-forward network there are one or more hidden layers that have no direct contact with the external layers. These networks strengthen the computational capability of the network (see Figure 6.2) [1–3].

In a feedback network if the output is directed back to the same layer or preceding layer nodes as input, then the feedback loop is created (see Figure 6.3). Recurrent networks are feedback networks with closed loops [1–3].

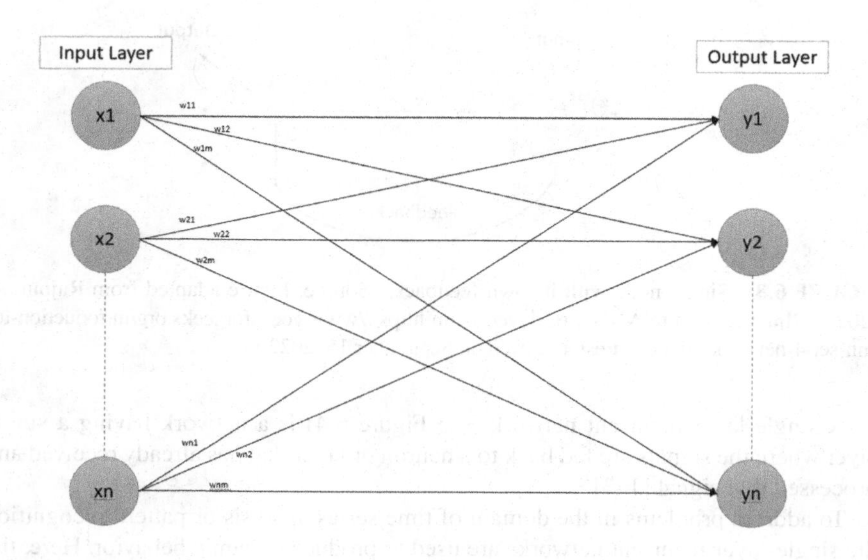

FIGURE 6.1 Single-layer feed-forward network. (Source: Figure adapted from Rajput, A. (2022). "Introduction to ANN,"retrieved from https://www.geeksforgeeks.org/introduction-to-ann-set-4-network-architectures/, retrieved on September 15, 2022.)

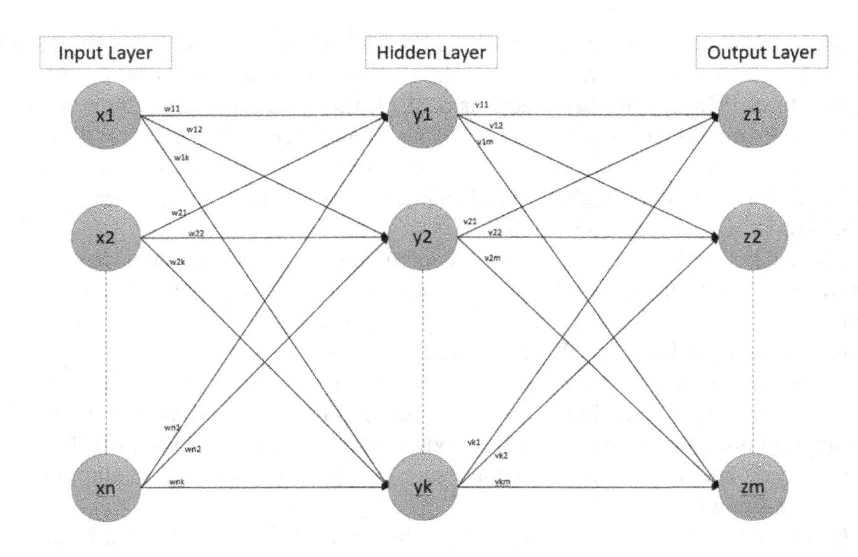

FIGURE 6.2 Multilayer feed-forward network. (Source: Figure adapted from Rajput, A. (2022). "Introduction to ANN", retrieved from https://www.geeksforgeeks.org/introduction-to-ann-set-4-network-architectures/, retrieved on September 15, 2022.)

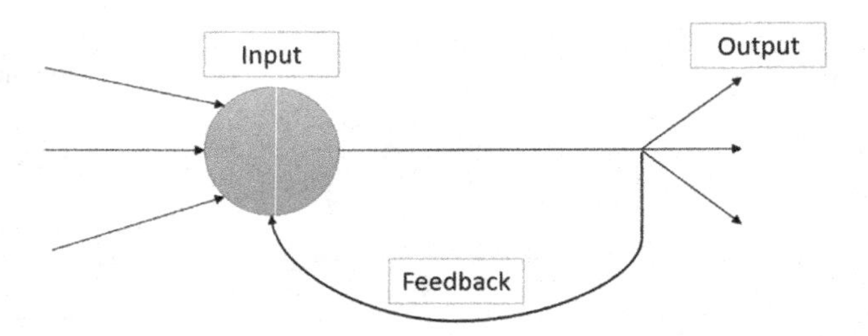

FIGURE 6.3 Single node with its own feedback. (Source: Figure adapted from Rajput, A. (2022). "Introduction to ANN", retrieved from https://www.geeksforgeeks.org/introduction-to-ann-set-4-network-architectures/, retrieved on September 15, 2022.)

A single-layer recurrent network (see Figure 6.4) is a network having a single layer where the signals are fed back to a neuron or layer that has already received and processed that signal [1–3].

To address problems in the domain of time series analysis or pattern recognition the single-layer recurrent networks are used to produce dynamic behavior. Here, the learning process is reinforced by using an internal memory.

The multilayer recurrent network has multiple recurrent layers that are applied on top of each other [1–3] (see Figure 6.5).

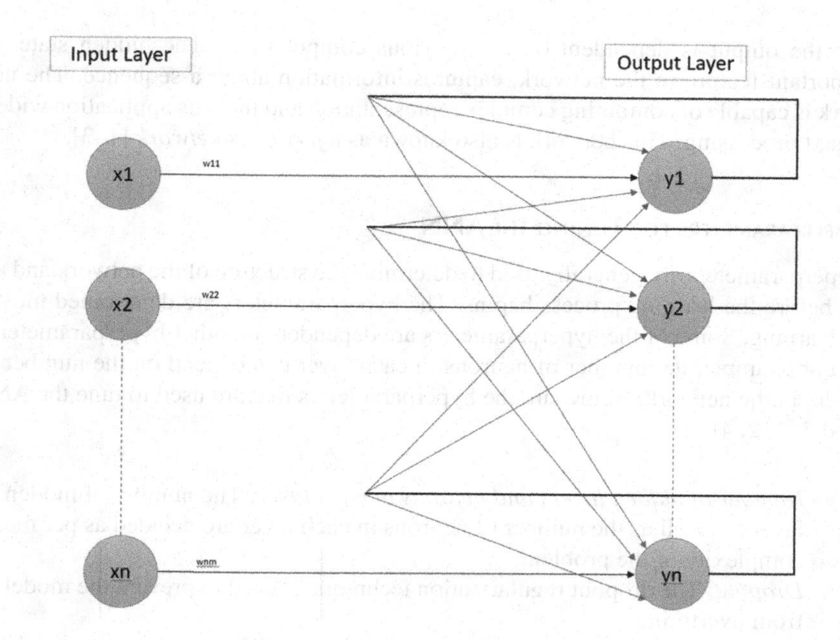

FIGURE 6.4 Single-layer recurrent network. (Source: Figure adapted from Rajput, A. (2022). "Introduction to ANN," retrieved from https://www.geeksforgeeks.org/introduction-to-ann-set-4-network-architectures/, retrieved on September 15, 2022.)

In this network, the same task is implemented on every element of the sequence

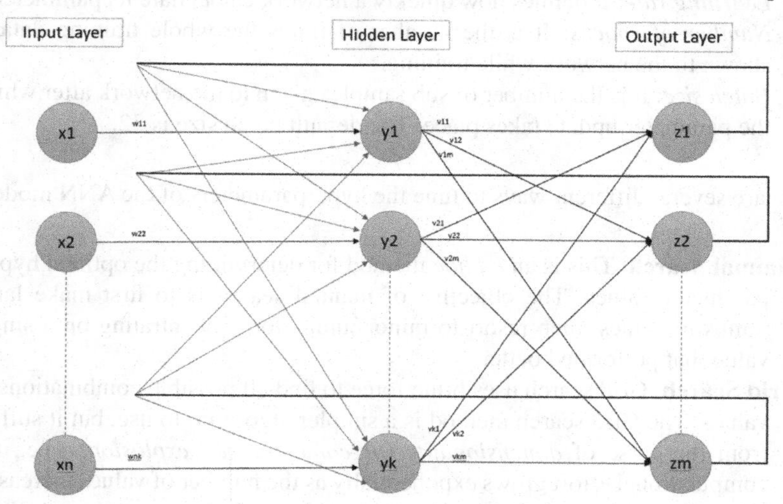

FIGURE 6.5 Multilayer recurrent network. (Source: Figure adapted from Rajput, A. (2022). "Introduction to ANN,"retrieved from https://www.geeksforgeeks.org/introduction-to-ann-set-4-network-architectures/, retrieved on September 15, 2022.)

and the output is dependent on the previous computations. The hidden state, an important feature in the network, captures information about a sequence. The network is capable of computing complex representation and finds its application widely in text processing. This network is also known as a *feedback network* [1–3].

Hyperparameters for Tuning the ANN

Hyperparameters are generally used to determine the structure of the network and are set before the learning process begins. The hyperparameters are determined mostly by learning. Some of the hyperparameters are dependent on other hyperparameters.

For example, the number of neurons in each layer can depend on the number of layers in the network. Below are the hyperparameters that are used to tune the ANN model [1, 2, 4]:

- *Number of hidden layers and neurons in each layer*: The number of hidden layers as well as the number of neurons in each layer are decided as per the complexity of the problem.
- *Dropout*: The dropout regularization technique is used to prevent the model from overfitting.
- *Network weight initialization*: It is better to use different network weight initialization techniques depending on the use of the activation function in different layers.
- *Activation function*: The activation function decides how to perform computation on the input signal to get a desired type of output signal.
- *Learning rate*: It defines how quickly a network can update its parameters.
- *Number of epochs*: It is the number of times the whole training data is shown to the network while training.
- *Batch size*: It is the number of sub samples given to the network after which the parameter update takes place. The default batch size is 32.

There are several different ways to tune the hyperparameters of the ANN model [4].

Manual Search: This is an *ad hoc* method for determining the optimal hyperparameter values. The objective of manual search is to first make large jumps in values, then resort to minor jumps for concentrating on a single value that performed better.

Grid Search: Grid search uses brute force to find all possible combinations of values. The Grid search method is a simpler algorithm to use, but it suffers from the curse of *dimensionality* (*or combinatorial explosion*)—i.e., the computational effort grows exponentially as the number of values increases.

Random Search: Random search, unlike grid search, uses a statistical distribution for each hyperparameter from which the values are randomly sampled. The model's hyperparameter values will be set for each iteration by sampling the stated distributions above. The optimization will be faster but less accurate.

An Example of ANN Classification

Let's use the IRIS dataset to demonstrate the modeling of an ANN classifier. The Iris Dataset contains four features (predictors) of certain flower images, i.e., the length and width of the sepals and petals across 150 samples (instances) [5]. In this dataset there are three species namely the *Setosa*, the *Virginica*, and the *Versicolor*. The measures of the features across each specie were used to create an ANN model that can learn the characteristics of each species and to classify them. This dataset is often used in data mining, classification, and clustering examples to test the algorithms.

Let's use the R scripting language to demonstrate the ANN model built to classify instances in the IRIS dataset. First let us load the IRIS dataset executing the following R script [5]

```
data(iris)
iris$setosa <- iris$Species == "setosa"
iris$virginica <- iris$Species == "virginica"
iris$versicolor <- iris$Species == "versicolor"
```

Now we split the Iris dataset into a training and validation dataset with 70% of the instances used for training the ANN model and 30% of the instances used for validating the ANN model [5]

```
iris.train.idx <- sample(x = nrow(iris), size =
  nrow(iris)*0.7)
iris.train <- iris[iris.train.idx,]
iris.valid <- iris[-iris.train.idx,]
```

To build the ANN model install the *neuralnet* package and load the package in R using the following commands [5]

```
Install.packages("neuralnet")
library(neuralnet)
```

Now let's create an ANN model with the species as the response variable. The sepal (length and width) and petal (length and width) both will be used as the predictors. The ANN model, i.e., MLP has two hidden layers with 10 neurons in each [5].

```
iris.net <- neuralnet(setosa+versicolor+virginica ~
                    Sepal.Length + Sepal.Width + Petal.Length +
                       Petal.Width,
                   data = iris.train, hidden = c(10,10), rep = 5,
                      err.fct = "ce",
                   linear.output = F, lifesign = "minimal",
                      stepmax = 1000000,
                   threshold = 0.001)
```

Finally, we display the results using the following R command [5]

```
plot(iris.net, rep="best")
```

The resultant ANN model fitted on the training dataset output is shown in Figure 6.6.

Now that the ANN model has been trained let's classify the instances in the validation dataset. We execute the following R scripts to obtain the confusion matrix on the validation dataset [5].

```
iris.prediction <- compute(iris.net, iris.valid[-5:-8])
idx <- apply(iris.prediction$net.result, 1, which.max)
predicted <- c('setosa', 'versicolor', 'virginica')[idx]
table(predicted, iris.valid$Species)
```

The resultant confusion matrix is shown in Figure 6.7.

From Figure 6.7 it is evident that only 2 instances from the *virginica* class have been misclassified into the *versicolor* class. Therefore, we can compute the overall accuracy of prediction in the validation dataset as $\frac{16+12+15}{16+12+17} = 95.5\%$ [5].

See Sidebar 2 for more discussion of the interpretation of the confusion matrix.

Next, we will focus our discussion on ensemble learning–based classifiers.

INTRODUCTION TO ENSEMBLE LEARNING TECHNIQUES

Ensemble learning distinguishes the strong learners from the weak learners. A *strong learner* is a classifier or a regressor, which has the capability to reach to the highest potential accuracy thus minimizing both the bias and the variance. This means that a strong learner is theoretically able to achieve a non-null probability of misclassification with a probability of equal or more than 50%. Generally, most of the machine learning tasks are normally strong learners even if their domains are very limited. For example, a logistic regression cannot solve non-linear problems [2].

Conversely, a *weak learner* is a model that is able to achieve an accuracy slightly higher than a random guess. These types of learners have low complexity, but they can be trained very quickly. However, such learners can never be used alone to solve complex problems. In some very particular and small regions of the training space, a weak learner could reach a low probability of misclassification, but in the wide space its performance is just about superior then a random guess [2].

Now that we have seen the distinction between the strong and weak learners, let's formally define the ensemble. An *ensemble* is a set of weak learners that are trained together (or in a sequence) to make up a team. Both in classification and regression type problems, the final result is obtained by averaging the predictions or by employing a majority vote of all the classifiers (weak learners) [1, 2].

The most common approaches to ensemble learning are as follows 1, 2, 6]:

Bagging or bootstrap aggregating: This approach trains n weak learners (decision trees) using n training sets created by randomly sampling the original dataset D. The sampling process is known as bootstrap sampling. This process is normally performed with replacement, so as to determine different data distributions. In addition, the weak learners are also initialized and trained using some degree of randomness. This ensures that the probability of having clones becomes very small and, at the same time, it's possible to

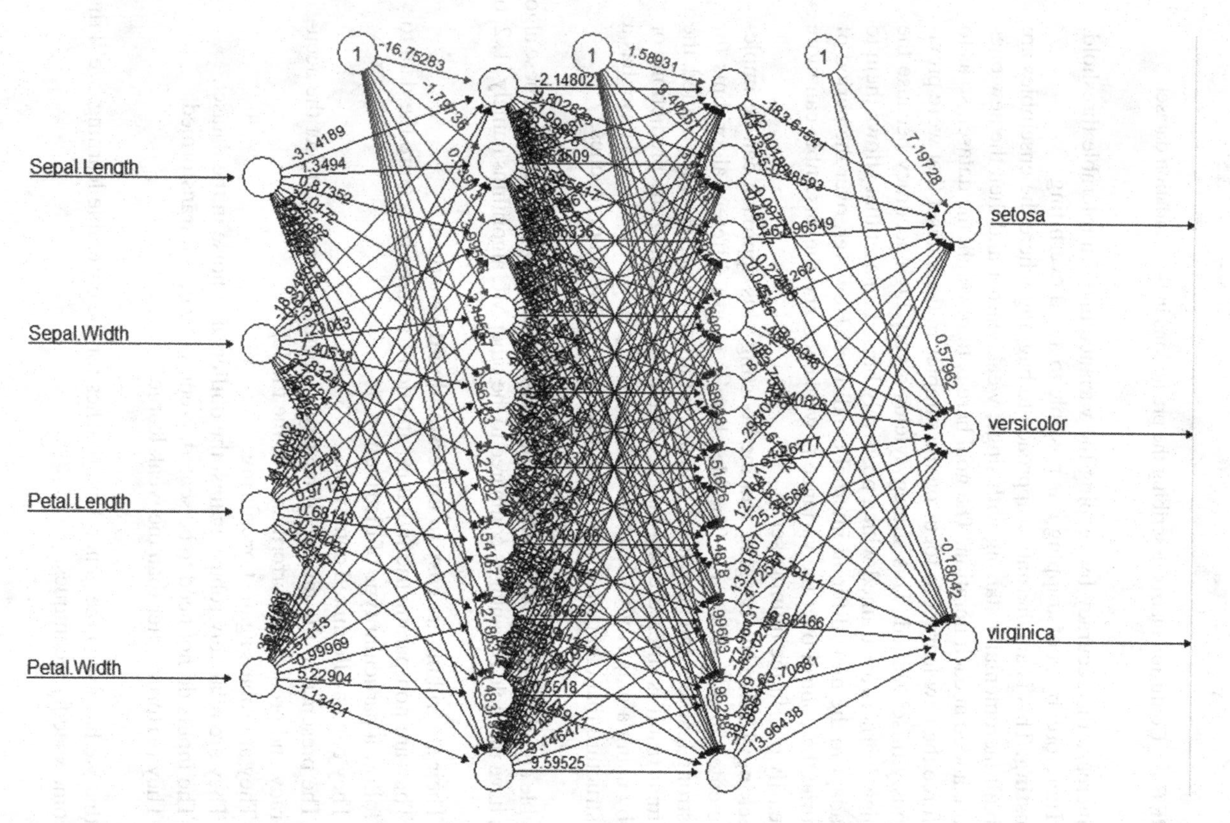

FIGURE 6.6 MLP containing 10 neurons in each of the hidden layer.

```
predicted    setosa versicolor virginica
  setosa         16          0         0
  versicolor      0         12         0
  virginica       0          2        15
```

FIGURE 6.7 Confusion matrix describing the predictions in the validation dataset.

increase the accuracy by keeping the variance under a tolerable threshold. Therefore, by bootstrapping it is possible to avoid overfitting.

Boosting: This is an alternative approach to bagging where the ensembles are built incrementally starting with single weak learner and then the new ones are added at each iteration. The goal here is to reweight the dataset, so as to force the new learner to focus on the samples in the dataset that were previously misclassified. This strategy yields a very high accuracy because the new learners are trained with a positively biased dataset that allows them to adapt to the most difficult internal conditions. However, over the period of time the control over the variance is weakened and the ensemble can more easily overfit the training set.

Stacking: Unlike the bagging and the boosting, this approach can be implemented in different ways. The idea here is to use different algorithms for strong learners and have them trained on the same dataset. Once trained, the final result is filtered using another classifier, averaging the predictions or by using a majority vote. This strategy is very powerful if the dataset has a structure that can be partially managed with different approaches.

Ensemble algorithms particularly utilize the decision trees as weak learners. These algorithms have multiple advantages compared to the other ML algorithms namely [1, 2, 6]

- Their algorithms are **easy to understand and visualize**.
- They are non-parametric in nature and don't assume or require the data to follow a particular type of distribution.
- They can handle mixed data types.
- The presence of the multi-collinearity of features does not affect the accuracy and prediction performance of the model.
- They are robust against overfitting.
- They are relatively robust against the outliers and noise in the dataset.
- The inputs do not need to be scaled, preprocessed, and transformed.
- They perform better than the weak learners.

Now that we have seen the general approaches for the ensemble learning, we'll discuss some specific techniques.

Random Forest Ensemble Learning

A Random Forest (RF) is a "bragging" ensemble model, which is, essentially, a collection of unpruned decision trees. This ensemble method tends to produce accurate models by reducing the instability found in single decision trees. In addition, the RFs are robust to changes in the training data and to noise in general. This feature is desirable when the dataset contains many outliers. Another advantage of the RF is that it has a built-in attribute selection technique, and its salient features demand very little preprocessing of the raw dataset [1, 2, 7].

We can summarize how the RF algorithm works as follows. RF builds many decision trees. Each tree is built from a random subset of data (from the training dataset) using the replacement strategy known as bagging and from a random subset of selected predictor variables. Each tree is built independently from one another receiving a vote and the majority rule applies here to determine the winner [1, 2, 7].

The RF produces a set of rules for classification, like that of any decision tree–based methods [7]. For RF several hyperparameters, including the number of trees (*ntree*), criterion of slitting (*Gini, Information Gain*), the maximum number of terminal nodes each tree in the forest can have (*maxnodes*), minimum size of terminal nodes (*nodesize*), and the number of variables randomly sampled as candidates at each split (*mtry*) can all be tuned either through random or through grid search to construct an optimal model. In a grid search, each axis of the grid is a parameter of an algorithm, and the points in the grid are specific combinations of hyperparameters.

Mathematically, the RF algorithm can be described as follows. Assume the data points of a sample $D=\{(x_1,y_1),\ldots,(x_n,y_n)\}$ are drawn randomly from a (possibly unknown) probability distribution $(x_i,y_i)\sim(X,Y)$. The goal is to build a classifier that can predict y from X based on the dataset of sample D. We are given the ensemble of weak classifiers $h=\{h_1(X),\ldots,h_k(X)\}$. If each $h_k(X)$ is a decision tree, then the ensemble is a RF. Let's define the parameters of the decision tree for the classifier $h_k(X)$ to be $\varnothing_k=\{\theta_{k1},\ldots,\theta_{kp}\}$. Thus, a decision tree k is a classifier $h_k(X)=h(X|\theta_k)$. The RF classifier can be defined as a family of classifiers $h(X|\theta_1),\ldots,h(X|\theta_k)$ based on a classification tree with parameters θ_k randomly chosen from a model random vector Θ [7].

Introduction to AdaBoost Ensemble Learning

Unlike the Random Forest, the AdaBoost (ADB) classifier is based on the principle of boosting. We have already discussed about the boosting approach. In summary, *boosting* is a general approach that can be applied to many statistical models. Boosting works in a sequential manner and does not involve bootstrap sampling. In ADB, each tree is fitted on a modified version of an original dataset. Finally, all the trees are added up to create a strong classifier. The working principle of the ADB classifier is summarized in Table 6.1 [1].

In step I the algorithm starts by fitting a simple classifier on the data. This so-called a *decision stump* splits the data into two regions. In step I, classes that are correctly classified are given less weight and the classes that are misclassified are given higher weight. In this iteration a new decision stump/weak classifier is fitted to the data and the weights are updated again in the subsequent iterations as discussed before. Once all the iterations are finished (step III), the results of the classification from each iteration are combined with their weights to produce a strong classifier, that predicts the classes [1].

TABLE 6.1
Working Principle of the ADB Algorithm

The algorithm for ADB consists of the following steps:

I. Initialize the observation weights $w_i = \frac{1}{N}, i = 1,2,\ldots,N$, where N = Number of observations.
II. For $m = 1$ to M:
Fit a classifier $GM(x)$ to the training data using weights w_i

Compute $err_m = \frac{\sum_{i=1}^{N} w_i I\left(y_i \neq G_m(x_i)\right)}{\sum_{i=1}^{N} w_i}$

Compute $\alpha_m = \log\left(\frac{1-err_m}{err_m}\right)$

Set $w_i \leftarrow w_i * \exp\left[\alpha_m * I\left(y_i \neq G_m(x_i)\right)\right], i = 1,2,\ldots,N.$

III. Now output: $G(x) = sign\left[\sum_{m=1}^{M} \alpha_m G_m(x)\right].$

The relevant **hyperparameters** for tuning the ABD are limited to the *maximum depth* of the weak learners/decision trees, the *learning rate*, and the *number of iterations/rounds*. The learning rate balances the influence of each decision tree on the overall algorithm, while the maximum depth ensures that samples are not memorized, but that the model will generalize well with the new data [6].

Introduction to Extreme Gradient Boosting (XGB)

The Extreme Gradient Boosting (XGB) algorithm is an ensemble learning method that is based on the gradient boosting principle. Gradient boosting works on the principle that the weak learners iteratively shift their focus toward problematic observations that were difficult to predict in the previous iterations. These ensembles of weak learners are typically decision trees. It builds the model in a stage-wise fashion like other boosting methods do, but it generalizes them by allowing optimization of an arbitrary differentiable loss function [8]. Table 6.2 discusses about the working principle of the gradient boosting learning method [1, 2, 8].

There are three elements involved in gradient boosting [1, 2, 8]

- The loss function, which depends on the type of problem being solved. In cases where classification problems are solved, the logarithmic loss is used. In boosting, at each stage, unexplained loss from prior iterations would be optimized rather than starting from scratch.
- Decision trees are used as a weak learner in gradient boosting.
- Trees are added one at a time and the existing trees in the model are not changed. The gradient descent procedure is used to minimize the loss while adding the trees.

TABLE 6.2
Working Principle of the Gradient Boosting Classifier

Initially, a model is fit on observations producing a certain accuracy $\left(F(x)\right)$ and the remaining unexplained variance is captured in the *error* term as shown below:

$$Y = F(x) + \text{error}$$

Then another model is fit on the error term to pull the extra explanatory component and add it to the original model (see the equation below), which should improve the overall accuracy:

$$Y = F(x) + G(x) + \text{error2, where error} = G(x) + \text{error2}$$

We can continue in this manner, i.e., fit a model on the **error2** component to extract a further explanatory component as :

$$\text{error2} = H(x) + \text{error3}$$

Now, model accuracy is further improved, and the final model equation is:

$$Y = F(x) + G(x) + H(x) + \text{error3}$$

Here, if we use weighted average (higher importance given to better models that predict results with greater accuracy than others) rather than simple addition, it will improve the results further. Therefore, the final model equation is:

$$Y = \alpha * F(x) + \beta * G(x) + \gamma * H(x) + \text{error3}$$

The steps involved in gradient boosting are

1. Initialize $f_0(x) = \text{argmin}_\gamma \sum_{i=1}^{N} L(y_i, \gamma)$
2. For $m = 1$ to M do the following

   For $i = 1,2,3,\ldots,N$ $r_{im} = -\left[\frac{\partial L(y_i, f(x_i))}{\partial f(x_i)}\right]_{f=f_{m-1}}$

   Fit a regression tree to the targets r_{im} giving terminal regions R_{im}, where $j = 1,2,3,\ldots,j_m$

   For $j = 1,2,3,\ldots,j_m$ compute $y_{jm} = \text{argmin}_\gamma \sum_{x_j \in R_{jm}} L\left(y_i, f_{m-1}(x_i) + \gamma\right)$

   Update $f_m(x) = f_{m-1}(x) + \sum_{j=1}^{j_m} \gamma_{jm} I(x \in R_{jm})$
3. Output $f'(x) = f_M(x)$

The hyperparameters that are available to tune the XGB classifier are *learning rate, column subsampling, regularization, subsample* (which is bootstrapping the training sample), *maximum depth* of trees, *minimum weights in child nodes for splitting* and the *number of estimators* (trees). These hyperparameters are frequently used to address the bias-variance-trade-off. While higher values for the number of estimators, regularization, and weights for the child nodes are associated with decreased overfitting, the learning rate, maximum depth, subsampling, and column subsampling all need to have lower values to reduce overfitting [1, 2, 6, 8].

CROSS-VALIDATION

Cross-validation is a popular technique that is used to determine the true performance of a classifier. Cross-validation also ensures robustness in the model. Cross-validation, however, carries a significant computational expense [1, 2].

Earlier it was mentioned that in the modeling methodology, a model is developed on the training dataset and evaluated on the test dataset. However, there are cases in which the train and the test dataset have not been selected homogeneously. In such cases there may be some unseen extreme cases appearing in the test dataset but not in the training dataset. This situation would result in degrading the performance of the model. To avoid this effect cross-validation should be performed [1, 2].

In cross-validation the dataset is divided into equal parts and the training of the model is performed on all the other parts of the dataset except for one in which the performance of the model is evaluated. This process is then repeated again and again until the model has been trained on the entire dataset [1, 2].

Let's consider an example. The 10-*fold cross-validation* technique is the most popular technique used for measuring the true performance of a classifier. In 10-fold cross-validation, the dataset is divided into ten parts, subsequently trained on nine parts, and tested on the remaining one part. This process is performed ten times, in order to cover all the parts of the data. Finally, the error calculated is averaged over all the errors.

Consider the *German Credit* dataset to illustrate the difference in the performance of the ensemble classifiers namely RF and XGB [9]. The *German Credit* data contains data on 20 variables and the classification of whether an applicant is considered a good or a bad credit risk for 1000 loan applicants. The predictors in this dataset can be mainly classified under the categories for the applicant's demographic and socio-economic profiles.

First let's install the appropriate packages in R using the following commands [9]

```
install.packages("randomForest", "ROCR", "gbm")
library(randomForest)
library(ROCR)
library(gbm)
```

The *randomForest* package provides access to build the RF classifier and the *gbm* package provides access to build the XGB classifier. The package *ROCR* provides functionality to measure the performance of the classifier [9].

Now let's access the *German Credit* dataset, preprocess it (convert variables of type characters to factors, and redefine the range of the response variable), and split the dataset into train and test sets in the ratio of 70% and 30%, respectively, using the following script [9]:

```
set.seed(5000)
#Reading Data
german_credit = read.table("http://archive.ics.uci.edu/ml/
  machine-learning-databases/statlog/german/german.data")
#Assigning variable names
```

```
colnames(german_credit)=c("chk_acct","duration","credit_
  his","purpose","amount","saving_acct","present_
  emp","installment_rate","sex","other_debtor","present_
  resid","property","age","other_install","housing","n_
  credits","job","n_people","telephone","foreign","respo
  nse")
#Response is in 1,2 - we need to change it to 0,1
german_credit$response = german_credit$response - 1

#Dividing into training and testing dataset
index <- sample(nrow(german_credit),size =
  nrow(german_credit)*0.70)
german_credit_train <- german_credit[index,]
german_credit_test <- german_credit[-index,]

german_credit_train[sapply(german_credit_train, is.character)]
  <- lapply(german_credit_train[sapply(german_credit_train,
  is.character)],as.factor)
german_credit_test[sapply(german_credit_test, is.character)]
  <- lapply(german_credit_test[sapply(german_credit_test,
  is.character)],as.factor)
```

Now we'll fit the RF classifier on the training dataset. The R command to fit the RF classifier is provided below (note the use of the hyperparameters here) [9].

```
credit.rf <- randomForest(as.factor(response)~., data =
  german_credit_train,mtry=sqrt(ncol(german_credit_train)-1),
  ntree=1000)
credit.rf
```

Figure 6.8 summarizes the output, i.e., the fitted RF model on the training dataset.

Now we'll plot the error versus tree graph to see how the errors vary with respect to the number of trees in the training dataset using the following script [9]:

```
plot(credit.rf, lwd=rep(2, 3))
legend("right", legend = c("OOB Error", "FPR", "FNR"),
  lwd=rep(2, 3), lty = c(1,2,3), col = c("black", "red",
  "green"))
```

```
Call:
 randomForest(formula = as.factor(response) ~ ., data = german_credit_train,
dit_train) - 1), ntree = 1000)
               Type of random forest: classification
                     Number of trees: 1000
No. of variables tried at each split: 4

        OOB estimate of  error rate: 24.86%
Confusion matrix:
    0  1 class.error
0 446 41  0.08418891
1 133 80  0.62441315
```

FIGURE 6.8 Fitted RF model on the training dataset.

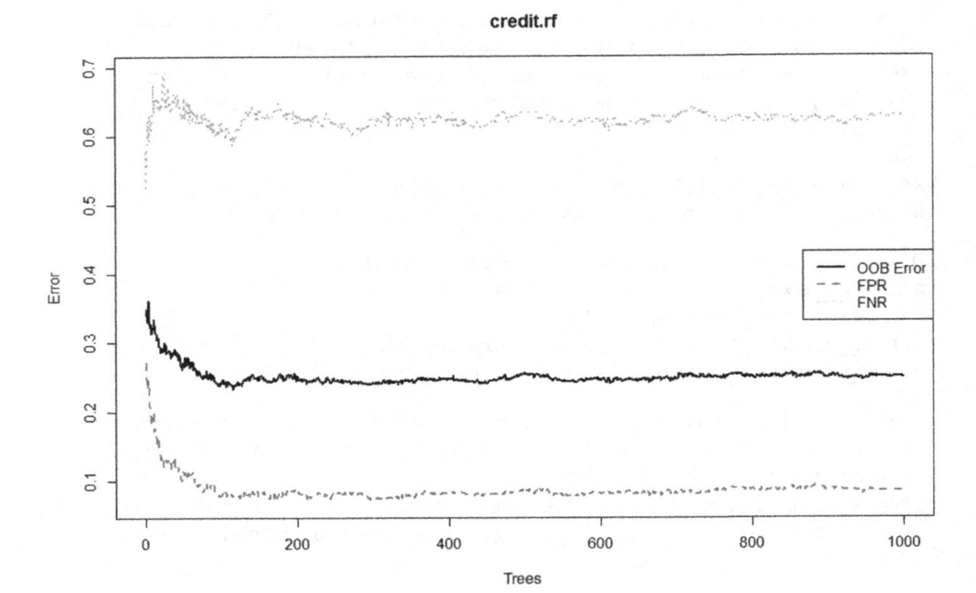

FIGURE 6.9 Comparison of the OOB, FPR, and FNR error. (Source: Figure adapted from Rathod, V. (2020). "German Credit Scoring Data." Retrieved from https://rpubs.com/vidhividhi/GermanCreditData, retrieved on August 6, 2022.)

The results are shown in Figure 6.9.

Now we'll determine the result of the prediction on the test dataset. First we determine the confusion matrix (see Sidebar 2) using the script:

```
## confusion matrix
credit.rf.pred_test<- predict(credit.rf,newdata=german_credit_
  test, type = "prob")[,2]
optimal.pcut= .16#our assumption
credit.rf.pred.class.test<- (credit.rf.pred_test>optimal.
  pcut)*1
table(german_credit_test$response, credit.rf.pred.class.test,
  dnn = c("True", "Pred"))
```

The results are shown in Figure 6.10.

Here it is evident that there is a significant amount (141) of false negatives, i.e., good applicants being mis-predicted as bad applicants. Complete details of the performance metrics for the confusion matrix are provided in Sidebar 2.

```
     Pred
True  0   1
   0 72 141
   1  5  82
```

FIGURE 6.10 Confusion matrix demonstrating the performance of the RF classifier on the test dataset.

The receiver operating characteristic (ROC) is an important measure of predictor performance (see Sidebar 2 for more details). The R script for obtaining the ROC curve for the German Credit dataset is:

```
#roc
pred <- prediction(credit.rf.pred_test,
   german_credit_test$response)
perf <- performance(pred, "tpr", "fpr")
plot(perf, colorize=TRUE)
```

The results of executing the script are giving in Figure 6.11.

This ROC curve suggests that the performance of the classifier is significantly better than random guessing.

Another important measure of predictor performance is the Area Under the Curve (AUC). Again, see Sidebar 2 for more details. Here, the AUC is 0.81 which can be obtained by executing the following R command [9]

```
unlist(slot(performance(pred, "auc"), "y.values"))
```

An AUC = .81 indicates a reasonably good separation of classes (discrimination) by the predictor.

Next, let's compare the performance of the RF classifier against the XGB classifier. Upon executing the R script provided below the XGB model is fitted on the training dataset [9].

```
credit.bo= gbm(response~., data = german_credit_train,
  distribution = "bernoulli",n.trees = 100, shrinkage = 0.01,
  interaction.depth = 8)
summary(credit.bo)
```

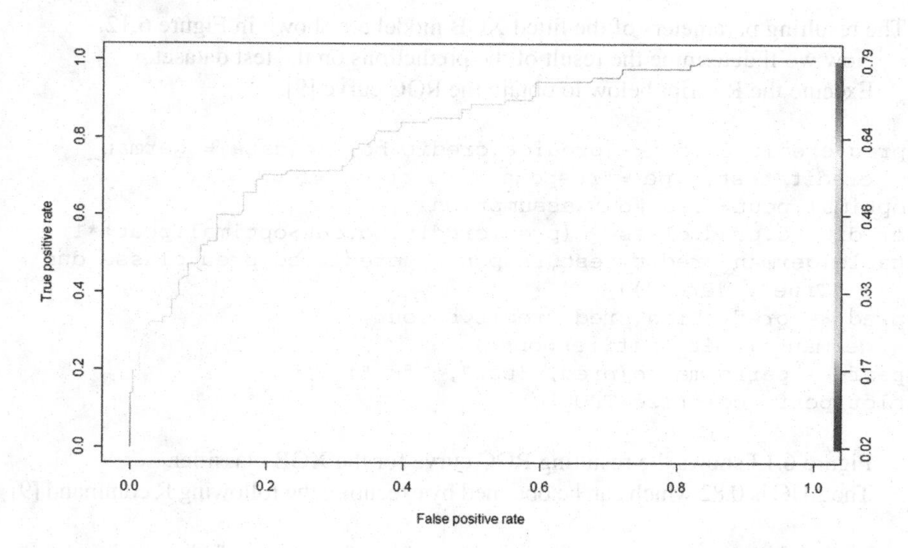

FIGURE 6.11 ROC curve demonstrating the performance of the RF classifier. (Source: Figure adapted from Rathod, V. (2020). "German Credit Scoring Data." Retrieved from https://rpubs.com/vidhividhi/GermanCreditData, retrieved on August 6, 2022.)

```
##                                   var     rel.inf
## chk_acct                     chk_acct 24.79473162
## purpose                       purpose 13.93340620
## duration                     duration 12.26373402
## amount                         amount  7.88413169
## other_install           other_install  6.58860998
## credit_his                 credit_his  5.69527536
## present_emp               present_emp  5.05194876
## saving_acct               saving_acct  4.90412351
## property                     property  4.55080555
## age                               age  4.49352334
## installment_rate installment_rate  1.93167495
## present_resid           present_resid  1.88800520
## housing                       housing  1.75965162
## job                               job  1.43809345
## sex                               sex  1.35982765
## other_debtor             other_debtor  0.63424947
## n_credits                   n_credits  0.34181637
## telephone                   telephone  0.26795525
## n_people                     n_people  0.13894866
## foreign                       foreign  0.07948735
```

FIGURE 6.12 Fitted XGB model on the training dataset. (Source: Figure adapted from Rathod, V. (2020). "German Credit Scoring Data." Retrieved from https://rpubs.com/vidhividhi/GermanCreditData, retrieved on August 6, 2022.)

The resulting parameters of the fitted XGB model are shown in Figure 6.12.

Now we'll determine the result of the predictions on the test dataset.

Execute the R script below to obtain the ROC curve [9].

```
pred.credit.bo.out<- predict(credit.bo, newdata = german_
  credit_test,type ="response" ,n.trees =100 )
optimal.pcut= .16 #our assumption
credit.bo.pred.class<- (pred.credit.bo.out>optimal.pcut)*1
table(german_credit_test$response, credit.bo.pred.class, dnn =
  c("True", "Pred"))
pred <- prediction(pred.credit.bo.out,
  german_credit_test$response)
perf <- performance(pred, "tpr", "fpr")
plot(perf, colorize=TRUE)
```

Figure 6.13 shows the resulting ROC curve for the XGB classifier.

The AUC is 0.82 which can be obtained by executing the following R command [9]

```
unlist(slot(performance(pred, "auc"), "y.values"))
```

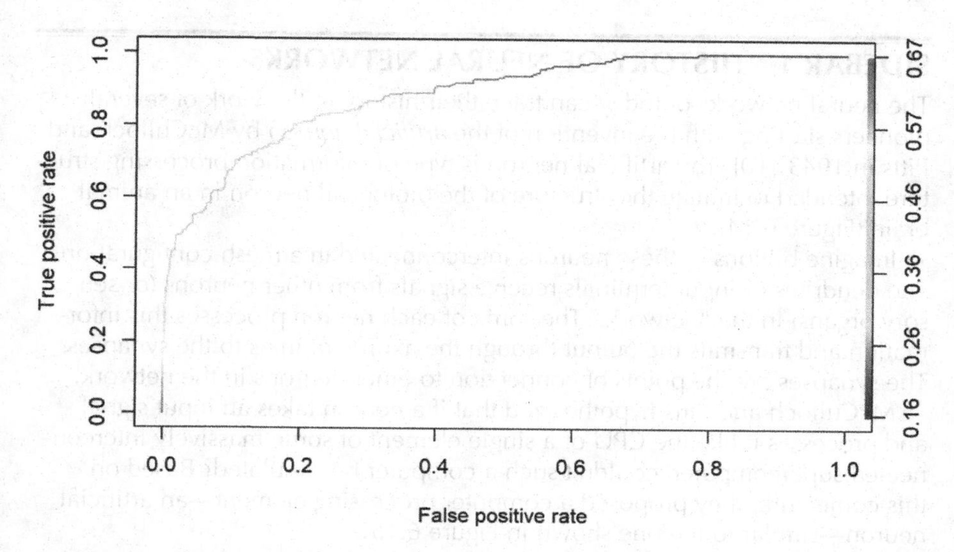

FIGURE 6.13 ROC curve demonstrating the performance of the XGB classifier. (Source: Figure adapted from Rathod, V. (2020). "German Credit Scoring Data." Retrieved from https://rpubs.com/vidhividhi/GermanCreditData, retrieved on August 6, 2022.)

Both the classifiers have a very similar performance. Although XGB performs slightly better than the RF (AUC = 0.82 versus 0.81).

Note: In order to perform 10-fold cross-validation during the training phase the script needs to add the parameter *fold = 10*. For example, if we want to fit the RF model on the training dataset with 10-fold cross-validation then the updated R script should be:

```
credit.rf <- randomForest(as.factor(response)~., data =
  german_credit_train, fold=10, mtry=sqrt(ncol(german_credit_
  train)-1), ntree=1000)
credit.rf
```

SUMMARY

The focus of this chapter was to introduce the concept of supervised learning. The ANN architecture discussed in this chapter lays the foundation for discussing about the deep learning techniques in latter chapters. This chapter also lays the foundation for the ensemble techniques that have proven to prevent the models from overfitting when compared to the traditional ML classifiers. Also note that the potential of the ensemble classifiers for classification purposes is best obtained by careful selection of the various hyperparameters.

SIDEBAR 1 HISTORY OF NEURAL NETWORKS

The neural networks of today can trace their history to the work of several pioneers starting with the invention of the *artificial neuron* by McCulloch and Pitts in 1943 [10]. The artificial neuron is type of information processing structure intended to imitate the structure of the biological neuron in an animal brain (Figure 6.14).

Imagine billions of these neurons interconnected in a mesh configuration. The dendrites or input terminals receive signals from other neurons (or sensory organs) in the "network." The soma of each neuron processes this information and transmits the output through the axon terminals to the synapses. The synapses are the points of connection to other neurons in the network.

McCulloch and Pitts hypothesized that if a neuron takes an input signal and processes it like the CPU of a single element of some massively interconnected supercomputer, couldn't such a computer be simulated? Based on this conjecture, they proposed a computer processing element—an artificial neuron—similar to the one shown in Figure 6.15.

Here the x_i takes the place of a dendrite and the summation operation simulates the soma and all $w_i=1$. $f(x)$, the decision function, is typically the sigmoid, hyperbolic tangent or heavyside step function, i.e., essentially a threshold detector.

In 1958, Frank Rosenblatt proposed a more general artificial neuron model called the *perceptron*, which improved on the McCulloch–Pitts neuron in that weights and thresholds can change over time, i.e., the system can learn [11].

The work of McCulloch, Pitts, Rosenblatt, and others spurred a frenzy of activity into massively scaled computing models based on simple processing elements. These included systolic and wavefront processors, transputers, and dataflow machines. But in 1969 Minsky and Papert published a book

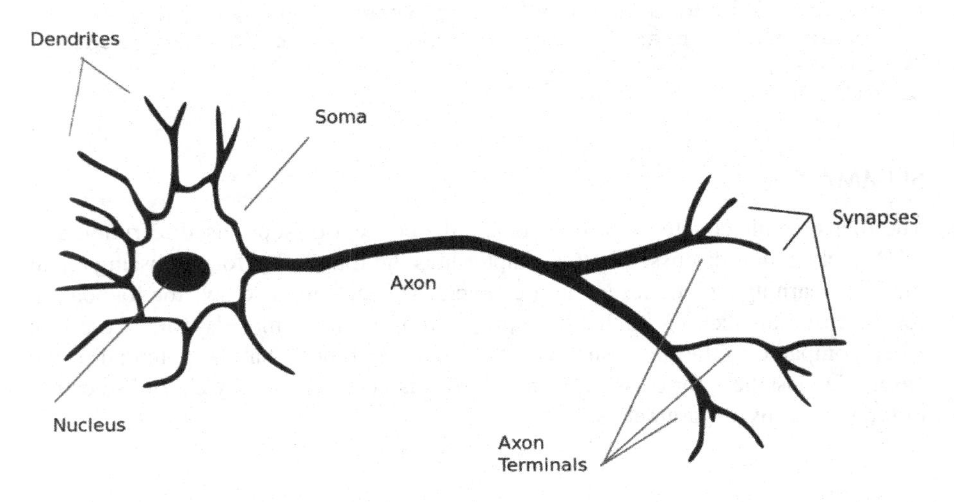

FIGURE 6.14 Model of an animal brain neuron.

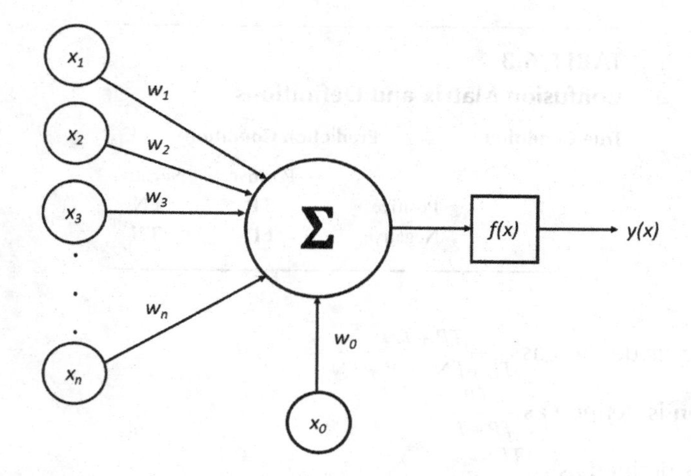

FIGURE 6.15 McCulloch and Pitts' artificial neuron (1943) [10].

criticizing the limitations of the perceptron. In addition, overhype of AI's potential, government research funding cutbacks, reduced interest in hard-wired connected computing and market factors led to a so-called "AI winter of the late 1970 and 1980s." During this time interest in neural networks (and other forms of AI) and massively parallel computing, was greatly reduced. Advances in computing processing power and the discovery of multilayer learning in the 1990s through the early 2000s led to an AI resurgence, highlighted by many of the theories and techniques discussed in this book.

SIDEBAR 2 CONFUSION MATRIX AND PERFORMANCE MEASURES

A confusion matrix (Table 6.3) is a matrix of the actual versus the predicted outcome [1].

True positives (TP): True positives are cases when both the true and the prediction conditions are positive.

True negatives (TN): True negatives are cases when both the true and the prediction conditions are negative.

False positives (FP): When we predict the condition as positive given that it is actually negative. FPs are also considered to be type I errors.

False negatives (FN): When we predict the condition as negative given that it is actually positive. FNs are also considered to be type II errors.

Now let's define some metrics based upon the four quadrants of the confusion matrix [1].

TABLE 6.3
Confusion Matrix and Definitions

True Condition	Prediction Condition	
	Positive	Negative
Positive	**TP**	**FN**
Negative	**FP**	**TN**

Accuracy is defined as $\frac{TP+TN}{TP+FN+FP+TN}$

Precision is defined as $\frac{TP}{TP+FP}$

Recall is defined as $\frac{TP}{TP+FN}$

Specificity is defined as $\frac{TN}{TN+FP}$

F1 score (F1): It is the harmonic mean of the precision and recall and is given as $\frac{2*\text{Precision}*\text{Recall}}{\text{Precision}+\text{Recall}}$

Sensitivity or the **True Positive Rate (TPR)** is given as $\frac{TP}{TP+FN}$

False Positive Rate (FPR) is given as $\frac{FP}{FP+TN}$

Receiver operating characteristic (ROC) curve is used to plot between the **TPR** and **FPR**. See Figure 6.16.

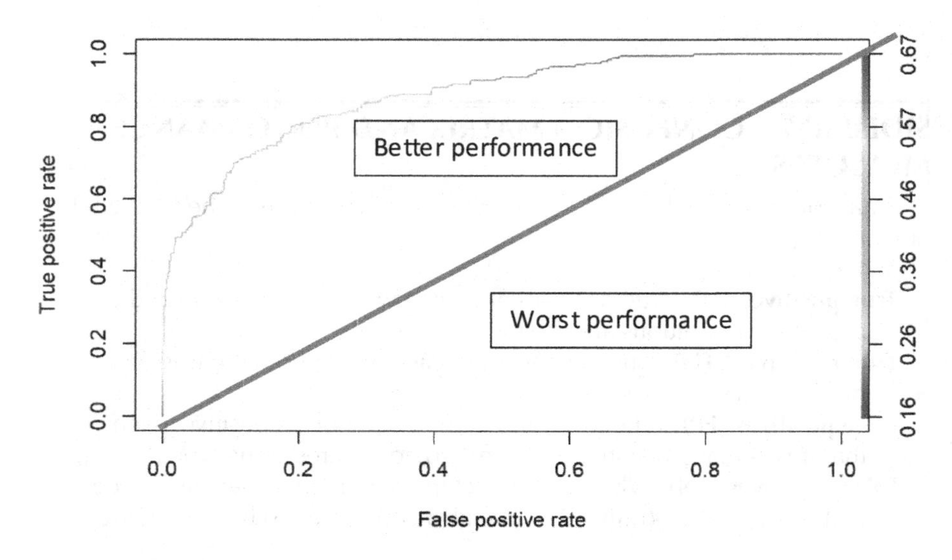

FIGURE 6.16 Receiver Operating Curve (ROC).

Note the bold line. The area below the dashed line suggests that the performance of the classifier is worse than a random guess. Conversely, the area above the dashed line suggests that the performance of the classifier is much better than a random guess. Ideals the best classifier is the one for which the TPR is always 1 across all the values of the FPR. The Area Under the Curve (AUC) metric indicates how successful a model is at separating positive and negative classes. AUC is obtained from the ROC curve. The higher the AUC the better the separation of classes [1].

EXERCISE

1. Download the Iris dataset (discussed in lesson 6) and extract a random sample containing 80% of the instances from this dataset. First, summarize this dataset. Second, use the ANN classifier to classify the different species using all the predictors. For validation testing, split the dataset into a ratio of 80:20. Clearly highlight the choices for the hyperparameters used in ANN modeling and summarize the performance of the classifier. To report the performance of the classifier show the confusion matrix, determine both the class-wise and overall accuracy, plot the ROC curve, and determine the AUC.

2. Using the Random Forest classifier, classify the species on the Iris dataset containing only 80% of the instances as obtained in the previous question. Discuss your choices for the use of hyperparameters. List the predictors used for the classification purpose. Does the model overfit? For validation testing, split the dataset into a ratio of 80:20. To report the performance of the classifier show the confusion matrix, determine both the class-wise and overall accuracy, plot the ROC curve, and determine the AUC.

3. Using the AdaBoost classifier, classify the species on the Iris dataset containing only 80% of the instances. Use the same dataset that you used for the questions 1 and 2. Discuss your choices for the use of hyperparameters. Compare the performance of the AdaBoost classifier against the Random Forest classifier. For validation testing, split the dataset into a ratio of 80:20. To report the performance of the classifier show the confusion matrix, determine both the class-wise and overall accuracy, plot the ROC curve, and determine the AUC.

4. Using the XGBoost classifier, classify the species on the Iris dataset containing only 80% of the instances. Use the same dataset that you used for the questions 1, 2, and 3. Discuss your choices for the use of hyperparameters. Compare the performance of the XGBoost classifier against the Random Forest and the AdaBoost classifier. For validation testing, split the dataset into a ratio of 80:20. To report the performance of the classifier show the confusion matrix, determine both the class-wise and overall accuracy, plot the ROC curve, and determine the AUC.

5. Repeat every instruction provided in questions 1 to 4. This time use the 10-fold cross-validation technique to report the performance of the classifier on the training dataset. Use the entire Iris dataset with 80% of the dataset as the training set.

6. What are the merits of using the ensemble learning algorithms compared to using the traditional machine learning classifiers for classification purposes?

REFERENCES

1. Dangeti, P. (2017). *Statistics for Machine Learning*. Packt Publishing, ISBN 978-1-78829-575-8.
2. Bonaccorso, C. (2018). *Mastering Machine Learning Algorithms*. Packt Publishing, ISBN 978-1-78862-111-3.
3. Rajput, A. (2022). "Introduction to ANN", retrieved from https://www.geeksforgeeks.org/introduction-to-ann-set-4-network-architectures/, retrieved on September 15, 2022.
4. Jordan, J. (2017). "Hyperparameter Tuning for Machine Learning Models", retrieved from https://www.jeremyjordan.me/hyperparameter-tuning/, retrieved on August 6, 2022.
5. Milton, V. (2017). "Iris- Neural Network", retrieved from https://rpubs.com/vitorhs/iris, retrieved on August 6, 2022.
6. Nikulski, J. (2020). "The Ultimate Guide to AdaBoost, Random Forestsand XGBoost", retrieved from https://towardsdatascience.com/the-ultimate-guide-to-adaboost-random-forests-and-xgboost-7f9327061c4f, retrieved on August 6, 2022.
7. Biau, G. (2012). Analysis of a Random Forests model. *Journal of Machine Learning Research*, 13, 1063–1095.
8. Chen, T., Guestrin, C. (2016). XGBoost: A Scalable Tree Boosting System. *Proceedings of the KDD*, 16, 1–10.
9. Rathod, V. (2020). "German Credit Scoring Data", retieved from https://rpubs.com/vidhividhi/GermanCreditData, retrieved on August 6, 2022.
10. McCulloch, W. S., and Pitts, W. (1943). A Logical Calculus of the Ideas immanent in Nervous Activity. *The Bulletin of Mathematical Biophysics*, 5(4), 115–133.
11. Rosenblatt, F. (1958). The Perceptron: A Probabilistic Model for Information Storage and Organization in the Brain. *Psychological Review*, 65(6), 386. Since the decision function depends linearly on the inputs x_i, the perceptron is a *linear classifier*.

7 Natural Language Processing for Analyzing Unstructured Data

Before discussing Natural Language Processing (NLP) tasks and techniques, it is important to understand its need for analytical purposes. The data available to us for any research or analysis endeavor can be either in a structured or unstructured format. Earlier chapters have focused on analyzing structured (i.e., tabular) data and in that process, we have learned how to uncover hidden information in the data by utilizing different analytical techniques.

In a similar manner, unstructured data (e.g., text, image, and audio) also contain hidden information that can be mined. For example, the social media site Twitter contains a corpus of tweets, each about 280 character long. A collection of these tweets from any individual holds a wealth of information on how that person communicates their thoughts, emotions (happiness, anxiety, depression, etc.), and feelings (positive and negative sentiments) within their social network. By performing sentiment and/or emotion classification on tweets, we can infer the thoughts and feelings of the individuals on any given topics. Intuitively, unstructured data are equally as valuable as structured data. For both preprocessing is needed before mining information and is perhaps more important with unstructured data. The types of analyses that can be performed on unstructured data, however, are very different from the techniques that are employed to mine structured data [1–6].

NLP is a subfield of Artificial Intelligence (AI) that assists computers in understanding the meaning of human language. By utilizing NLP, AI-based machines can help process information from data, which can help humans in gaining significant insights. As technology advances and massive amounts of text data are generated every day, the need to access and process it becomes increasingly more significant [1, 6].

Human language, also called *natural languages*, are diverse and rich in their beauty and expressive powers. Natural languages can consist of words and a set of rules for organizing those rules to form meaning. We call the rules set for any particular natural language as *grammar*. Using the structure of grammar to analyze human speech or writing is possible to an extent and forms the basis for any text or verbal chatbot. But the science here is inexact. Humans never completely follow grammatical or syntactic rules in conversation, and not even the stodgiest professor follows these rules in formal writing. Moreover, human speakers and writers use local dialects or slang and may have accents, which further complicates any kind of structured analysis of language [1, 6].

DOI: 10.1201/9781003278177-7

We use NLP techniques to perform analysis and process massive volumes of text data across the digital world including social media, search engines, online reviews, news reports, blogs, etc. Examples of NLP applications include [1, 6]:

- Building spam filter classifiers.
- Web-based search engines that provide information retrieval services (e.g., Google search).
- Machine translation from one natural language to another (e.g., Google translate).
- Q&A: automatic response to emails or chats.
- Summarizing documents.
- Sentiment analysis and emotion classification.
- Speech processing: options on phone helplines.
- Optical character recognition to scan cheques at ATMs.
- Spell check, autocomplete facilities in Google search, etc.

In any case, the study of analysis of conversation (or writing) for the purpose of human/computer interaction is beyond the scope of this text. Rather, we will focus on analyzing speech and writing for the purposes of extracting insight from that information. We'll begin with some terminology used in NLP.

TERMINOLOGY FOR NLP

The language data that all NLP tasks depend upon is called the text corpus. A *corpus* (from the Latin for "body") is a large collection of text data that can be in any natural language. The corpus can consist of a single document or a set of documents. The source of the text corpus can be social network sites such as Twitter, Facebook, blog sites, open discussion forums, etc. In audio data, words, phonemes, or other sounds of interest form the corpus used in NLP. The forgoing discussion, however, is focused on textual data only [1, 6].

A paragraph is the largest unit of text handled by an NLP task. Paragraphs can be further broken down into sentences. Sentences are composed of words which are formed using letters or alphabets in a given language. Tokenizers can be used to split a document into paragraphs, paragraphs into sentences, sentences into words, and words into characters.

Sentences are the next level of lexical unit in a language data. A sentence encapsulates a complete meaning or thought and context. It is usually extracted from a paragraph based on boundaries determined by punctuations such as the period or question mark. The sentence also conveys opinion or sentiment and emotions expressed in it. It is important to note that sentences consist of parts of speech (POS) entities like nouns, verbs, adjectives, and so on. Tokenizers are available to split paragraphs to sentences based on punctuations.

Phrases are a group of consecutive words within a sentence that can convey a specific meaning. Some of the NLP tasks extract key phrases from sentences for search and retrieval applications.

The next smallest unit of text is the word. The common tokenizers split sentences into words based on punctuations like spaces and comma [1, 6]. One of the problems with NLP is the ambiguity in the meaning of same words used in different contexts.

Finally, the smallest unit of text are the characters. Tokenizers are available or can be designed to split words into characters which when combined together form the word. While the single character data is not usually interesting, the frequency of two-letter combinations (e.g., aa, ab, ac, …., zy, zz) in writing samples can represent an authorship "fingerprint." That is, each writer tends to have a unique statistical distribution of use of these two-letter combinations. Two-letter combination analysis has been used in plagiarism detection and in the identification of authorship for anonymous writing [1, 6]. For example, two-word analysis has been used with great success to help identify the anonymous authors of the *Federalist Papers*.

There are two commonly used language data on which NLP tasks are performed. They are *n*-grams and bag-of-words (BOW). A sequence of characters or words forms an *n*-gram. For example, a character unigram consists of a single character, a bigram consists of a sequence of two characters, and so on. Similarly, word *n*-grams consists of a sequence of *n* words. In NLP, *n*-grams are used as features for text classification [1, 6].

BOW in contrast to *n*-grams does not consider word order or sequence. It captures the word occurrence frequencies in the text corpus. BOW is also used as features in tasks like sentiment analysis, emotion classification, and topic identification. Later in this chapter, we will discuss the BOW and *n*-grams in detail.

In order to demonstrate the NLP tasks, we will have to first install the NLTK and its associated modules in Python.

INSTALLING NLTK AND OTHER LIBRARIES

The basic concepts and tasks of NLP will be demonstrated using the appropriate packages in Python. Before we can install these packages, you have to make sure that the latest version of PIP is installed in your system. To install PIP in your computer, refer to this link https://pip.pypa.io/en/stable/installation/.

Next, we'll explore NLP in Python starting with the NLTK (natural language toolkit) and NUMPY (numerical Python) packages. To install NLTK and NUMPY type the following commands at the command prompt [1, 6]

```
pip install -U nltk
pip install -U numpy
```

The output of executing the NLTK installation commands are shown in Figure 7.1. The output for NUMPY will be similar.

After installing the NLTK and NUMPY navigate to the Python prompt and type the following commands [1, 6]

```
import nltk
nltk.download()
```

You should be able to see the NLTK Downloader GUI navigator window from Githubusercontent.com (Figure 7.2).

Now click on the tab "*All Packages*" and double click on the packages that needs to be installed. The output is shown in Figure 7.3.

```
C:\Users\sus64>pip install -U nltk
Collecting nltk
  Downloading https://files.pythonhosted.org/packages/6f/ed/9c755d357d33bc1931e157f537721efb5b88d2c583fe593cc09603076cc3
/nltk-3.4.zip (1.4MB)
    100% |████████████████████████████████| 1.4MB 4.7MB/s
Requirement not upgraded as not directly required: six in c:\users\sus64\appdata\local\programs\python\python35\lib\site
-packages (from nltk) (1.11.0)
Collecting singledispatch (from nltk)
  Downloading https://files.pythonhosted.org/packages/c5/10/369f50bcd4621b263927b0a1519987a04383d4a98fb10438042ad410cf88
/singledispatch-3.4.0.3-py2.py3-none-any.whl
Building wheels for collected packages: nltk
  Running setup.py bdist_wheel for nltk ... done
  Stored in directory: C:\Users\sus64\AppData\Local\pip\Cache\wheels\4b\c8\24\b2343664bcceb7147efeb21c0b23703a05b23fcfea
ceaa2a1e
Successfully built nltk
Installing collected packages: singledispatch, nltk
Successfully installed nltk-3.4 singledispatch-3.4.0.3
You are using pip version 10.0.1, however version 19.0.1 is available.
```

FIGURE 7.1 Installing NLTK package in Python.

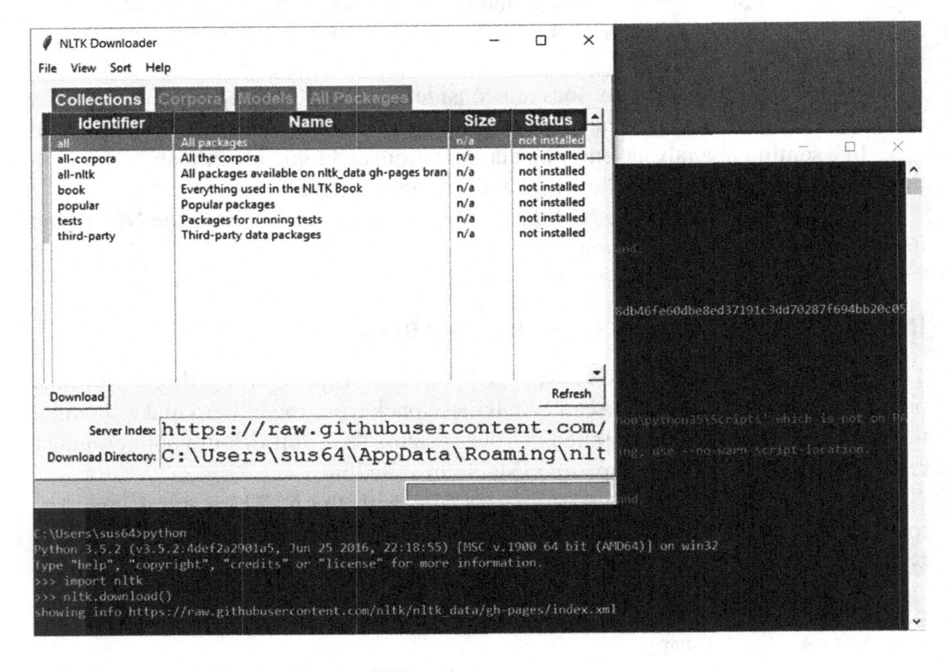

FIGURE 7.2 NLTK downloader GUI navigator.

The packages listed in Table 7.1 should be installed for the purpose of completing the exercises discussed in this chapter. Now let us discuss some basic NLP tasks.

TOKENIZATION

Tokenization refers to the identification of words and special characters and symbols from text. Tokenization of text is one of the basic NLP tasks [1, 6]. Let's perform tokenization on the following tweet.

@BOMBSKARE the anti-Scottish feeling is largely a product of Tory press scare-mongering. In practice most people won't give a toss!

NLTK Downloader

File View Sort Help

Collections Corpora Models All Packages

Identifier	Name	Size	Status
abc	Australian Broadcasting Commission 2006	1.4 MB	not installed
alpino	Alpino Dutch Treebank	2.7 MB	not installed
averaged_perceptron_	Averaged Perceptron Tagger	2.4 MB	installed
averaged_perceptron_	Averaged Perceptron Tagger (Russian)	8.2 MB	not installed
basque_grammars	Grammars for Basque	4.6 KB	installed
biocreative_ppi	BioCreAtIvE (Critical Assessment of Information Ex	218.3 KB	not installed
bllip_wsj_no_aux	BLLIP Parser: WSJ Model	23.4 MB	not installed
book_grammars	Grammars from NLTK Book	8.9 KB	not installed
brown	Brown Corpus	3.2 MB	installed
brown_tei	Brown Corpus (TEI XML Version)	8.3 MB	installed
cess_cat	CESS-CAT Treebank	5.1 MB	not installed
cess_esp	CESS-ESP Treebank	2.1 MB	not installed
chat80	Chat-80 Data Files	18.8 KB	not installed
city_database	City Database	1.7 KB	not installed
cmudict	The Carnegie Mellon Pronouncing Dictionary (0.6)	875.1 KB	not installed
comparative_sentenc	Comparative Sentence Dataset	272.6 KB	not installed

Download Refresh

Server Index: https://raw.githubusercontent.com/

Download Directory: C:\Users\sus64\AppData\Roaming\nlt

Finished installing brown_tei

FIGURE 7.3 Downloading packages from the NLTK downloader GUI navigator. (Note: At the bottom of the NLTK downloader GUI navigator you should see a message that will indicate the status of the installation of the packages. The status bar on the bottom right will show you how much of the task has been completed at any given time.)

TABLE 7.1
List of Packages to be Installed

Identifier	Description
Brown	Brown Text Corpus
Gutenberg	Gutenberg Text Corpus
Twitter_samples	Twitter messages sample
Universal_tagset	Universal POS tag mapping
Webtext	Web text corpus
stopwords	Stopwords Corpus

To do the tokenization, execute the following Python code [1, 6]:

```
# Perform tokenization on the sample tweets
# first import the nltk package
import nltk
```

```
# from the nltk corpus import twitter_samples and rename it as
  ts
from nltk.corpus import twitter_samples as ts
# show the contents of the corpus
ts.fileids()
# Now let us get a sample of tweets from the corpus in the
  json format
samples_tw = ts.strings('tweets.20150430-223406.json')
# let us consider a sample tweet
samples_tw[20]
# Now let us tokenize a sample tweet
# import the in-built word tokenizer and rename it as wtoken
from nltk.tokenize import word_tokenize as wtoken
wtoken(samples_tw[20])
```

Upon executing the above script, the tokens obtained from the tweet are

```
['@', 'BOMBSKARE', 'the', 'anti-Scottish', 'feeling',
  'is', 'largely', 'a', 'product', 'of', 'Tory', 'press',
  'scaremongering', '.', 'In', 'practice', 'most', 'people',
  'wo', "n't", 'give', 'a', 'toss', '!']
```

Note here that there are few special characters including @ and !. Later we will see an example of tokenization that will demonstrate how to remove unwanted special characters [1, 6].

In order to split the tweet based on punctuation the NLTK package provides a *wordpunct* tokentzer [1, 6].

```
from nltk import wordpunct_tokenize
wordpunct_tokenize(samples_tw[20])
```

Upon executing the script on the tweet text as before, the tokens obtained are

```
['@', 'BOMBSKARE', 'the', 'anti', '-', 'Scottish', 'feeling',
  'is', 'largely', 'a', 'product', 'of', 'Tory', 'press',
  'scaremongering', '.', 'In', 'practice', 'most', 'people',
  'won', "'", 't', 'give', 'a', 'toss', '!']
```

Note here that the word "*anti-scottish*" is separated into three tokens namely *anti*, -, and *Scottish* [1, 6].

A *regular expression* is a simplified way to represent character strings that can be used to match qualifying text. The Unix/Linux operating systems support regular expressions representation for use in commands such as grep, sed, awk, and so on. NLTK also provides a regular expression feature called RegEx pattern. It is very easy to build customized tokenizers using the RegEx pattern. For example, consider the following commands [1, 6]:

```
from nltk import regexp_tokenize
patn = '\w+'
regexp_tokenize(samples_tw[20],patn)
```

Here the specified RegEx pattern is *'\w+'*, which removes special characters. Upon executing the above script, the tokens obtained from the tweet are

```
['BOMBSKARE', 'the', 'anti', 'Scottish', 'feeling', 'is',
  'largely', 'a', 'product', 'of', 'Tory', 'press',
  'scaremongering', 'In', 'practice', 'most', 'people', 'won',
  't', 'give', 'a', 'toss']
```

This is the most desirable tokenized output [6].

Now let's try a different RegEx pattern, namely the pattern *'\w+|[!,\-,]'* using the command script [1, 6]:

```
patn = '\w+|[!,\-,]'
regexp_tokenize(samples_tw[20],patn)
```

Upon executing the above script, the tokens obtained from the tweet are

```
['BOMBSKARE', 'the', 'anti', '-', 'Scottish', 'feeling',
  'is', 'largely', 'a', 'product', 'of', 'Tory', 'press',
  'scaremongering', 'In', 'practice', 'most', 'people', 'won',
  't', 'give', 'a', 'toss', '!']
```

Comparing the above output to the output obtained from the previous use of the RegEx pattern, it can be noted that the punctuation characters have reappeared (but not the special character @). There are many other kinds of RegEx patterns that can be used to preprocess text in useful ways [1, 6].

STEMMING

Stemming is a text preprocessing task for transforming related or similar variants of a word (such as *talking*) to its base form (to *talk*). Stemming is important because with many words more than one variant can share the same meaning (e.g., *talking* and *talk; walk* and *walking*) [1, 6].

The most basic stemming action is to reduce a plural word to its singular form. For example, *apples* can be reduced to *apple*. The stemmer utilizes the Porter algorithm which is basically a collection of language-specific rules (in this case, English) to derive the stem words. For example, a language-specific rule can be to remove suffixes such as "*ing*" from the word [1, 6].

Using the Python script shown below let's perform a few stemming actions [6]

```
# Import the NLTK package and the PorterStemmer package for
  stemming
import nltk
from nltk.stem import PorterStemmer
# Create a stemming instance
stemming = PorterStemmer()
# Now let us perform stemming
stemming.stem("talking")
stemming.stem("talks")
```

Upon executing the last two lines of the script, you will see the same output, i.e., *talk*.

Custom stemmers can be created using regular expressions as shown below [6]:

```
#Create a RegEx stemmer that removes 'able' or 'ing' with 4 as
  the minimum length of string to stem
from nltk.stem import RegexpStemmer
regexp_stemmer = RegexpStemmer("ing$|s$|e$|able"ing$|s$|e$|abl
  e$"#x0022;,min=4)
regexp_stemmer.stem("flyable")
regexp_stemmer.stem("flying")
```

Upon executing the last two lines of the script, you will see the same output, i.e., *fly*. Any substrings that match the regular expression will be removed. The variable *min* = 4 specifies the minimum length of the string to stem [6].

STOPWORDS

More commonly used words in English such as *the*, *is*, *he*, and so on, are generally called *stopwords*. Removing the stopwords is a common preprocessing step in an NLP application. Generally, words that do not signify any importance to the document, such as the articles (e.g., *the*, *a*, *an*) and pronouns (e.g., *he*, *she*, *they*) are removed during this preprocessing step [1, 6].

Using the Python script provided below let's see how the preprocessing step of stopwords removal is performed [6].

```
# Import the stopwords package from nltk.corpus
from nltk.corpus import stopwords
# Use the English stopwords
sw_l = stopwords.words('english')
# show the stopwords from the "English" language
sw_l[20:40]
```

The above script will result in an output showing a subset of the stopwords in the English language.

```
['himself', 'she', "she's", 'her', 'hers', 'herself', 'it',
  "it's", 'its', 'itself', 'they', 'them', 'their', 'theirs',
  'themselves', 'what', 'which', 'who', 'whom', 'this']
```

Let's see a simple example to illustrate the stopwords removal step in NLP. Consider the following Python script [6]:

```
# Remove stopwords from an example sentence
example_text = "This is an example sentence to test stopwords"
example_text_without_stopwords=[word for word in example_text.
  split()
if word not in sw_l]
# Display the resulting text after the removal of the stop words
example_text_without_stopwords
```

The script above outputs the resulting tokens that are not part of the stopwords list, which is

```
['This', 'example', 'sentence', 'test', 'stopwords']
```

Note that the token or word "*This*" was not removed in this case because character T is in uppercase. Therefore, it is important to first convert the entire text to lowercase characters and then perform the preprocessing step of stopwords removal. It should also be noted that the NLTK package provides stopwords corpora for 21 different languages [1, 6].

PART OF SPEECH TAGGING

Part of speech (POS) tagging is another important preprocessing step used for categorizing the words in a sentence into specific syntactic or grammatical functions. In the English language, the main parts of speech are *nouns*, *pronouns*, *adjectives*, *verbs*, *adverbs*, *prepositions*, *determiners*, and *conjunctions* [1, 6].

In the POS tagging task the objective is to first tokenize each word in a sentence, and then assign them to one of the main parts of speech categories. The NLTK provides both a set of tagged text corpus and a set of POS trainers for creating custom taggers. The most common tagged datasets in NLTK are the **Penn Treebank** and the **Brown Corpus** [1, 6]. Let's consider an example to illustrate how the POS tagging works.

```
# Import the nltk package and perform tokenization
import nltk
text1 = nltk.word_tokenize("I left the room")
text2 = nltk.word_tokenize("Left of the room")
# Now let us perform universal POS tagging
nltk.pos_tag(text1,tagset='universal')
nltk.pos_tag(text2, tagset='universal')
```

Upon executing the above Python script the output corresponding to *text1* and *text2* are

```
[('I', 'PRON'), ('left', 'VERB'), ('the', 'DET'), ('room',
  'NOUN')]
[('Left', 'NOUN'), ('of', 'ADP'), ('the', 'DET'), ('room',
  'NOUN')]
```

Here it is important to note that the token *left* in *text1* is tagged or categorized as a *verb* and in *text2* it is categorized as a *noun*. This is because in *text1* the word *left* specifies an action taken and in *text2* the word *left* refers to the location [6].

Now that we have learned the basic NLP tasks, let's explore a specific text corpus to see how the basic NLP tasks can be combined together to create an application. The objective of the application here is to determine the percentage of stopwords in a given text corpus [6].

Let's compute the percentage of stopwords in Shakespeare's play *Hamlet* obtained from the Gutenberg package in NLTK, using the following Python script [6]:

```
# import the gutenburg corpus and extract all the words
from nltk.corpus import gutenberg
words_in_hamlet = gutenberg.words('shakespeare-hamlet.txt')
# Now get all the stopwords for the English language
from nltk.corpus import stopwords
sw = stopwords.words('english')
# Retrieve all the words from the gutenburg corpus after
  removing the stopwords
words_in_hamlet_without_sw = [word for word in words_in_hamlet
  if word
not in sw]
# Now compute the percentage of actual words in the gutenburg
  corpus
len(words_in_hamlet_without_sw)*100.0/len(words_in_hamlet)
```

Upon executing this script, you should be able to see that 69.26% of the words in *Hamlet* corpus are not stop words which means that 30.74% (100–69.26) of the words in this corpus are stop words.

Now let's look at BOW and *n*-grams, both of which have the potential to features for NLP analysis techniques such as text mining and classification.

BAG-OF-WORDS (BOW)

BOW is a method of extracting features from text. It is a representation of text describing the occurrence of words within a document. Using the BOW concept one can convert variable-length texts into fixed-length vectors by counting the frequency of occurrences of each word. In this method, any information about the order in which the word appears within the document is ignored [1, 6].

In order to obtain the BOW for a corpus, the steps are to first to obtain all the unique words from each sentence in the corpus, then create a vocabulary and, finally, construct the fixed length vectors for each unit of text (i.e., sentence) within the corpus [1, 6]. The BOW method has many applications (see Sidebar 3).

Here's a simple example to illustrate the concept of BOW. Consider the corpus containing the first few lines of text from the Charles Dicken's book *A Tale of Two Cities* shown below

It was the best of times,

it was the worst of times,

it was the age of wisdom,

it was the age of foolishness,

The vocabulary for the above lines can be described as (see Table 7.2).

Now the BOW of the corpus can be seen in Table 7.3.

TABLE 7.2
Vocabulary of the Corpus

Vocabulary	Frequency
It	4
Was	4
The	4
Best	1
Of	4
Times	2
Worst	1
Age	2
Wisdom	1
Foolishness	1

TABLE 7.3
BOW Representation of the Corpus

Corpus	BOW Fixed-Length Vectors
	[It, Was, The, Best, Of, Times, Worst, Age, Wisdom, Foolishness]
It was the best of times	[1, 1, 1, 1, 1, 1, 0, 0, 0, 0]
it was the worst of times	[1, 1, 1, 0, 1, 1, 1, 0, 0, 0]
it was the age of wisdom	[1, 1, 1, 0, 1, 0, 0, 1, 1, 0]
it was the age of foolishness	[1, 1, 1, 0, 1, 0, 0, 1, 0, 1]

n-GRAMS

n-grams are simply a sequence of *n* words. Here, each word or token is called a *gram*. Let's consider a simple example to illustrate the concept of *n*-grams. For the sentence within the corpus

It was the best of times

The 1-grams are *It*, *was*, *the*, *best*, *of*, and *times*. Here, note that the vocabulary of the 1-gram and the BOW are the same. On the other hand, a 2-gram (*bigram*) is a two-word sequence of words such as "*It was*," "*was the*," "*the best*," "*best of*," and "*of times*." In a similar manner, a 3-gram (*trigram*) is a three-word sequence of words such as "*It was the*," "*was the best*," "*the best of*," and "*best of times*." Unlike BOW, the order of the appearance of the words in a sentence of the document (corpus) is tracked in the case of *n*-gram [1, 6].

Now that we have introduced the basic tasks and the features for NLP, let's discuss the important applications of NLP, namely sentiment and emotion classification.

TABLE 7.4
Examples of Emotional and Sentimental Tweets

Emotion	Sentiment	Tweet
Happy	Positive	Woot I finally got an iPod that I have been planning to buy forever
Sadness	Negative	It certainly is a sin and a shame...
Anger	Negative	My sister and I have had several vicious scarring fights this summer
Joy	Positive	I was pleased as punch to see my old friend

SENTIMENT AND EMOTION CLASSIFICATION

The advent of social media and microblogging sites has allowed individuals and communities to freely express their opinions, feelings, and thoughts on a variety of topics using short and limited size texts. Twitter is a well-known social media site that allows individuals to post short messages (*a.k.a. tweets*) with a 280-character limitation. In aggregate, these tweets hold a wealth of information on how an individual communicates their thoughts, emotions (happiness, anger, disgust, anxiety, depression, etc.) and sentiments (positive, negative) within their social network. By analyzing the tweets not only the emotion of an individual but the emotions of a larger group can be identified. The more commonly expressed state of emotions/sentiments and feelings include *anger, disgust, fear, joy, love, sadness, surprise, tensed, positive*, *negative*, etc. Table 7.4 provides examples of tweets that express different types of emotions and sentiments [2, 3, 7].

However, *Emotion* and *Sentiment classification*, i.e., determining the emotions and sentiments within a statement (sentence), document or a corpus is a very challenging task [2, 3].

There are four major challenges associated with emotion/sentiment classification of social media data [2, 3]:

1. Peculiar structure and size,
2. Large amount of data,
3. Labeling needed for classification, and
4. Discrimination between emotions.

Let's briefly elaborate these challenges.

First, unlike conventional texts, tweets are peculiar in terms of their structure and size, i.e., they are restricted to a length of 280 characters. In addition, the language used by people in tweets to express their emotions and sentiments is very different from the language used in other digitized documents like blogs, articles, and news [2, 3].

Second, the availability of features in the tweets is very large. Each tweet, when presented as a vector of features, exponentially increases the size of the available features. Refer back to the discussion on the BOW concept. The corpus would contain millions of features for a given topic. This exponential increase in the number of features would severely challenge the computational capabilities of the algorithms used for emotion and sentiment classification [2, 3].

The third major challenge is inherent to the characteristics of the algorithm or the classifier. For example, supervised techniques need labeled data for training the classifier. Due to the large volume of Twitter messages, it would be time-consuming and tedious to manually annotate them with emotion and sentiment classes and later use them for training the classifiers [1–3, 6].

The fourth challenge is that the inherent nature of the different types of emotions makes it very difficult to differentiate between them. According to the Circumplex model, there are 28 affect words or emotions. Few emotions are clustered so close to each other that it becomes very hard to differentiate between them (see Sidebar 1) [8]. For example, emotions such as *anger*, *tense*, and *alarmed* or *excited*, *delighted*, and *aroused* are very similar to each other [8].

For performing emotion and sentiment classification we will explore a well-known lexicon-based classifier, popularly known as NRC, so named because it is based on the National Research Council of Canada's affect lexicon and the NLTK library's WordNet synonym sets. Using NRC, Mohammad and Turney have put together emotion annotations for about 14,182 words (lexicons) by crowdsourcing to Amazon's *Mechanical Turk*. This lexicon is commonly referred to as the "*NRC* emotion association lexicon" or *EmoLex*. *EmoLex* has annotations for eight emotions: *anger*, *anticipation*, *disgust*, *fear*, *joy*, *sadness*, *surprise*, *trust* and two sentiments: *negative* and *positive*. This lexicon corpus was constructed based on the measures of *Strength of Association* (SOA) and *Pointwise Mutual Information* (PMI) (see Sidebar 2) [2–5].

Let's take some tweets as examples and try to determine the emotion and sentiment within it using the NRC classifier. Here, we will exploit the implementation of the NRC classifier in the *syuzhet* package using the R scripting language [5].

```
# First install and load the syuzhet package in R
install.packages("syuzhet")
library(syuzhet)
# Let us consider several tweets to perform sentiment and
  emotion classification
example_text <- "This was the best summer I have ever
  experienced.
I had a blast in california hanging out with my family and
  friends.
So many lies about who you're talking to, where you're going,
  what you're doing.
I miss you so much and can't wait to see Ginny, Sister friend
  and my mom in a couple months!!!
They are amazing.
I get a jolt of something REAL loud and it makes me jump.
She is afraid to.
This smile made me more scared.
He first tore up the toy car.
Markets do not measure everything accurately, and I find this
  assumption to be disgusting."
# Create a character object to get the NRC sentiments
S_E_C <- get_sentences(example_text)
# Now let us get the sentiments and emotions
nrc_output <- get_nrc_sentiment(S_E_C)
nrc_output
```

	anger	anticipation	disgust	fear	joy	sadness	surprise	trust	negative	positive
1	0	0	0	0	0	0	0	1	0	1
2	2	0	1	2	0	1	1	0	2	0
3	0	0	0	0	0	0	0	0	0	0
4	0	1	0	0	1	0	0	1	1	1
5	0	0	0	0	0	0	0	0	0	0
6	0	0	0	0	1	0	1	1	0	2
7	0	0	0	1	0	0	0	0	1	0
8	0	0	0	0	1	0	1	1	0	1
9	0	0	0	0	0	0	0	0	0	0
10	1	0	1	1	0	0	0	1	1	0

FIGURE 7.4 Emotion and sentiment expressed within tweets.

```
[1] "I had a blast in california hanging out with my family and friends."
[2] "Markets do not measure everything accurately, and I find this assumption to be disgusting."
```

FIGURE 7.5 Tweets that express the *anger* emotion.

Upon executing the above R script, you should be able to see an output similar to Figure 7.4. Note that the 4th tweet here "*I miss you so much and can't wait to see Ginny*" expresses the emotions of *joy*, *anticipation*, *trust*, and a *positive* sentiment. Here, it is important to note that there can be more than one emotion expressed within a tweet.

Also note the values indicated for each emotion class across different rows. If an emotion type is more intensely expressed within the sentence, then the cell corresponding to that column (emotion class) will hold a higher value. The following scripts can be used to determine which tweets express the *anger* or the *joy* emotion (see Figure 7.5 and 7.6) [5].

```
# Now let us pick those sentences that expresses the anger
  emotion
angry_items <- which(nrc_output$anger > 0)
S_E_C[angry_items]

# Now let us pick those sentences that expresses the joy
  emotion
joy_items <- which(nrc_output$joy > 0)
S_E_C[joy_items]
```

While NRC is an important emotion classifier, a major drawback is that it fails to determine the emotions and sentiments expressed in a sentence if none of its constituent words matches a known lexicon [5]. For more information about the NRC classifier and its implementation in R visit the link https://cran.r-project.org/web/packages/syuzhet/vignettes/syuzhet-vignette.html.

```
[1] "I miss you so much and can't wait to see Ginny, Sister friend and my mom in a couple months!!!"
[2] "I get a jolt of something REAL loud and it makes me jump."
[3] "This smile made me more scared."
```

FIGURE 7.6 Tweets that express the *joy* emotion.

SUMMARY

In this chapter, we introduced the terminology associated with modern natural language processing. We then discussed natural language processing techniques for text (and images) including BOW and n-grams and discussed many important applications such as spam filtering and in artificial intelligence, such as sentiment analysis. We then explored the use of the NRC for emotion classification. These applications are only a small subset of the many uses of these simple but powerful techniques discussed.

SIDEBAR 1 CIRCUMPLEX MODEL AND 28 DIFFERENT EMOTIONS

The Circumplex model is a very popular model of human emotions, which characterizes emotions through two dimensions: valence and arousal [3, 8]. The Circumplex model suggests that emotions can be distributed in a two-dimensional circular space with the vertical axis representing the *arousal* and the horizontal axis representing the *valence*. In the circular space, the center of the circle represents a neutral valence and arousal. All the emotional states can be represented with any level of the valence and arousal [8]. The Figure 7.7 pictorially represents the two-dimensional circular space of the

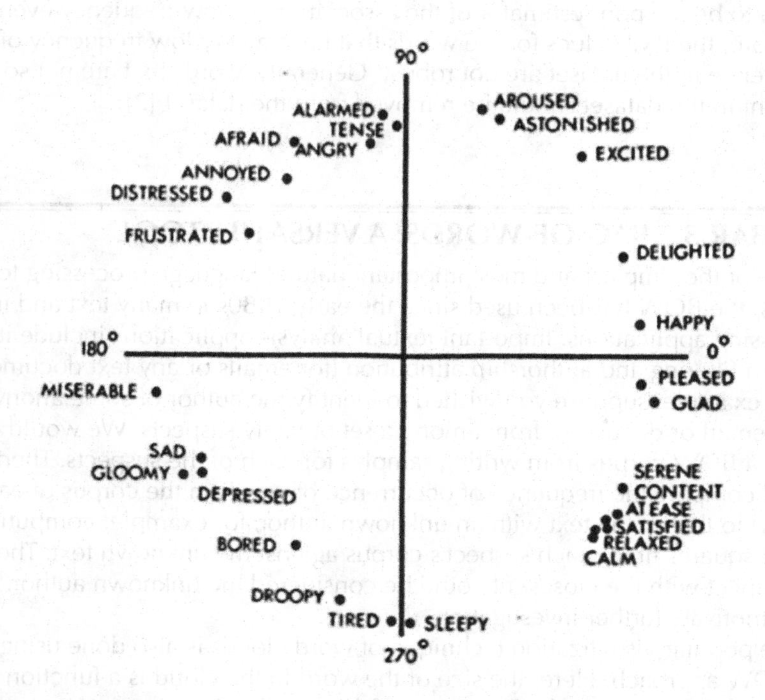

FIGURE 7.7 Two-dimensional Circumplex model for 28 affect words. (Source: Figure adapted from Russell, J. A. (1980). A circumplex model of affect. Journal of Personality and Social Psychology, 39(6), 1161–1178. https://doi.org/10.1037/h0077714.)

Circumplex model in which the positions of the 28 affect words or emotions are indicated. In Figure 7.7, the *happy* emotion is at an angle of 7.8° from the positive x-axis and represents a slight increase in arousal. Exactly at 90° is the emotion *surprise* which represents zero valence. Beyond 90° is a situation that is best characterized as involving negative valence and decrease in arousal, i.e., by the emotion *anger*. Beyond 180° is a state of negative arousal and negative valence which represents *sadness* [8].

SIDEBAR 2 PMI AND SOA

For each word a PMI determines the association of a word with a category (class). At the same time a secondary PMI, i.e., PMI′ is computed for each word that determines the association of a word with the other or remaining class labels. Finally, for each word the SOA is computed across each class which is the difference of PMI and PMI′. If a word has a stronger tendency to occur in a sentence with a class *X*, than in a sentence that does not belong to class *X*, then that word will have a SOA score greater than *zero* for class *X*. Such words are associated with the class *X*. These words are considered as potential lexicons for class *X*. The PMI measure on the other hand is well-known to be the poor estimator of the association for low-frequency events. Therefore, the PMI values for the words that have a very low frequency of occurrence in the dataset are not robust. Generally, words that are not so frequent in the dataset should be removed from the dataset [2].

SIDEBAR 3 BAG-OF-WORDS: A VERSATILE TOOL

As one of the simplest and most important natural language processing techniques, the BOW has been used since the early 1980s in many text and image processing applications. Important textual analysis applications include its use in spam filtering and authorship attribution (for emails or any text document).

For example, suppose you wished to identify the author of some anonymous email or document from among a set of likely suspects. We would create a BOW corpus from writing samples for each of the suspects. Then we would compare the frequency of occurrence of words in the corpus of each suspect to that of the text with an unknown author, for example, computing a least-squares fit for each suspect's corpus against the unknown text. Then the suspect with the closest fit could be considered the unknown author, or at least, motivate further investigation.

The popular visualization technique of word clouds is also done using the BOW approach. Here, the size of the word in the cloud is a function of the relative number of occurrences of the word in the corpus. Words appearing less than some minimum number of times are sometimes omitted as certain words such as infinitives. Colors are often used for certain word

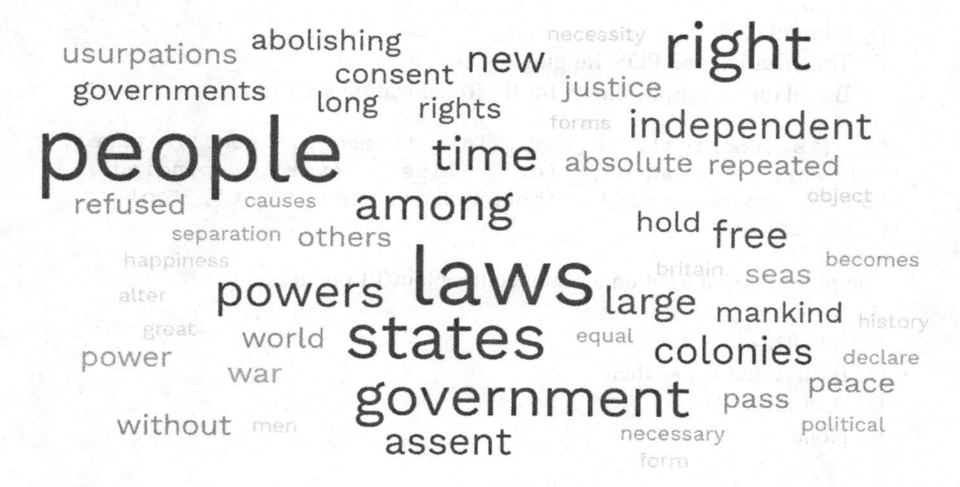

FIGURE 7.8 Word cloud for the U.S. Declaration of Independence, generated with Free Wordcloud Generator https://www.freewordcloudgenerator.com/

relationships or types. As an example, a wordcloud for the U.S. Declaration of Independence is shown in Figure 7.8.

In that document the word "people" appears 10 times, "powers" 5 times, and "alter" 2 times. Words appearing 1 time only are omitted.

The BOW approach has also been used for at least 30 years in image analysis by treating certain image features as "words" and building an associated corpus. The statistical properties of the corpus can then be used in applications ranging from image classification, texture analysis, and the identification of objects in moving models.

The simplicity and relative computational efficiency of BOW analysis means that it is likely to be used for a long time and will see many new applications in the future.

EXERCISE

1. One of the following is a TRUE statement
   A. N-grams does not consider word order or sequence
   B. The smallest unit of text is a sentence.
   C. Tokenization is one of the basic NLP tasks
   D. Both A and B are TRUE
   E. None are TRUE

2. A text preprocessing task that transforms related or similar variants of a word to its base form is coined as
   A. Tokenization
   B. Stop words removal
   C. POS Tagging

D. Stemming
E. Tokenization and POS Tagging

Based on the output shown for the Tokenization set T1:

```
T1: ['suttonnick', 'Friday, 's', 'Times', 'front', 'page',
'Miliband', 'savaged', 'for', 'lies', 'over', 'spending',
'tomorrowspaperstoday', 'bbcpapers', 'http', 't', 'co',
'ts9ZnULDwr']
```

3. The regex pattern used on a sentence to obtain T1 could be
   A. '\w+'
   B. '\w+|[!,\-,]'
   C. Both A and B together
   D. A or B
   E. None

4. From the set T1 we can infer that
   A. Stemming was performed before tokenization
   B. When POST tagged all the items will be *Noun*
   C. Before tokenization stop words were not removed
   D. Stemming was not performed before tokenization
   E. Both C and D

   Based on the sentiment and emotion classification by NRC provided below

   Code:

```
Install.packages("syuzhet")
 library(syuzhet)
 my_example_text <- "Further complicating the issue is
  Trump's warning that, if a bipartisan panel doesn't come
  up with a workable solution by February 15, he'll either
  shut down the government again—something Mitch McConnell
  and other Republicans on Capitol Hill have suggested they
  could overrule—or declare a national emergency to fund the
  wall without Congress. Putting the nuclear option on the
  table, as Trump apparently did against Kushner's wishes,
  may already have doomed a potential compromise. After all,
  what motivation does Trump's base have to go along with a
  middle ground when they're convinced, he can easily avoid
  compromising altogether?"
 s_v <- get_sentences(my_example_text)
 class(s_v)
 nrc_data <- get_nrc_sentiment(s_v)
 nrc_data
 fear_items <- which(nrc_data$fear > 0)
 s_v[fear_items]
 trust_items <- which(nrc_data$trust > 0)
 s_v[trust_items]
```

Output:

```
  anger anticipation disgust fear joy sadness surprise trust negative positive
1     0            0       1    3   0       1        1     1        2        1
2     0            0       0    1   0       1        1     0        1        1
3     0            0       0    1   0       0        0     3        1        0
```

5. The sum of all the valence, i.e., the sum of the difference between the sum of positive sentiment and sum of negative sentiment, in “my_example_text” is
   A. –2
   B. –1
   C. 0
   D. 6
   E. –6

6. The s_v[fear_items] will print
   A. The 1st sentence
   B. The 1st and 2nd sentence
   C. All the sentences
   D. Only the 3rd sentence
   E. NULL

7. The variable “s_v[trust_items]” will result in one of the following outputs
   A. After all, what motivation does Trump’s base have to go along with a middle ground when they’re convinced, he can easily avoid compromising altogether?
   B. Putting the nuclear option on the table, as Trump apparently did against Kushner’s wishes, may already have doomed a potential compromise.
   C. After all, what motivation does Trump’s base have to go along with a middle ground when they’re convinced, he can easily avoid compromising altogether? Putting the nuclear option on the table, as Trump apparently did against Kushner’s wishes, may already have doomed a potential compromise.
   D. Putting the nuclear option on the table, as Trump apparently did against Kushner’s wishes, may already have doomed a potential compromise. After all, what motivation does Trump’s base have to go along with a middle ground when they’re convinced, he can easily avoid compromising altogether?
   E. NULL

8. The lexicon corpus for NRC was constructed based on the measures of
   A. SOA
   B. PMI
   C. SOA and PMI
   D. N-grams
   E. Bag-of-words

9. Read the article "How emergency powers could be used to build Trump's wall" from the link https://www.bbc.com/news/world-us-canada-46784315. Your task is to collect textual instances (from any one or more sources, for example, Twitter, blogs, news articles, etc.) and discuss about the sentiment and emotions expressed by the US citizens toward Trump's plan to declare a state of emergency.

10. NRC is a
    A. Supervised classifier
    B. Unsupervised classifier
    C. Tokenizer
    D. Parser
    E. Lexicon-based classifier

REFERENCES

1. Arumugam R., Shanmugamani, R., (2018). *Hands-On Natural Language Processing with Python*, Packt Publishing, ISBN 978-1-78913-949-5.
2. Mohammad, S., Turney, P. (2011). Emotions Evoked by Common Words and Phrases: Using Mechanical Turk to Create an Emotion Lexicon. *Proceedings of the NAACL-HLT 2010 Workshop on Computational Approaches to Analysis and Generation of Emotion in Text*.
3. Hasan, M., Rundensteiner, E., Agu, E. (2014). EMOTEX: Detecting Emotions in Twitter Messages. *Academy of Science and Engineering, Bigdata/Socialcom/Cybersecurity Conference*, 27–31.
4. "NRC Word-Emotion Association Lexicon", retrieved from https://saifmohammad.com/WebPages/NRC-Emotion-Lexicon.htm, retrieved on May 23, 2022.
5. "Introduction to Syuzhet package", retrieved from https://cran.r-project.org/web/packages/syuzhet/vignettes/syuzhet-vignette.html, retrieved on May 23, 2022.
6. Dangeti, P. (2017). *Statistics for Machine Learning*. Packt Publishing Ltd., ISBN 978-1-78829-575-8.
7. Chapman, C., Feit, E. M. (2015). *R for Marketing Research and Analytics*, Springer, ISBN 978-3-319-14436-8.
8. Russell, J.A. (1980). A Circumplex Model of Affect. *Journal of Personality and Social Psychology*, 39, 1161–1178.

8 Predictive Analytics Using Deep Neural Networks

INTRODUCTION TO DEEP LEARNING

Deep Learning (DL) is a subfield of Machine Learning (ML), which, in turn, is a subfield of Artificial Intelligence (AI) [1–4] (Figure 8.1).

Before discussing DL, let us first more formally define AI and ML. AI is basically the art of enabling machines to perform tasks that require intelligence when performed by humans. AI encompasses many aspects including computer vision, knowledge reasoning, language processing, artificial agents, etc., and supporting mathematics, such as fuzzy logic for dealing with uncertain information (Sidebar 1). ML, on the other hand, comprises the collection of algorithms that learns patterns from the data. Finally, DL is the subset of ML that uses ANN to mimic how the brain works [1–4].

Let's consider a simple example to illustrate the concepts of AI, ML, and DL. Self-driving cars are an application of AI. One of the critical features of the self-driving cars is to recognize the boundaries of the road and other vehicles that may be obstacles, and to look out for pedestrians, cyclists, and so on. These functions require ML because it is not possible for machines to learn the necessary patterns (for a car, person, siderail, etc.) until the data (in most cases, images) is adequately provided. Finally, DL may be chosen as the method to implement this ML task which is to enable machines to learn the patterns (obstacles, road boundaries, etc.) [2, 3].

DL uses multiple layers to map the relationship between the input and the output. As previously discussed, each layer is a collection of neurons that perform a mathematical operation on its input. In DL, the "deep" architecture means that the model is large enough to handle multiple variables to approximate the patterns in the data. DL can also identify selective features that are crucial for the overall learning process, thus serving as a technique for dimensionality reduction. DL has proven particularly effective in the fields of image recognition, speech recognition, and NLP [2, 3]. It has completely transformed the ways in which images, text, and speech data can be used for learning patterns. The next section will introduce deep neural networks (DNN) and its architectural variants in detail.

THE DEEP NEURAL NETWORKS AND ITS ARCHITECTURAL VARIANTS

A *deep neural network* can be defined as a neural network with multiple hidden layers. To improve the prediction results using an ANN it is not enough just to keep on adding more nodes with a fixed (small) number of layers. Therefore, a DNN is

DOI: 10.1201/9781003278177-8

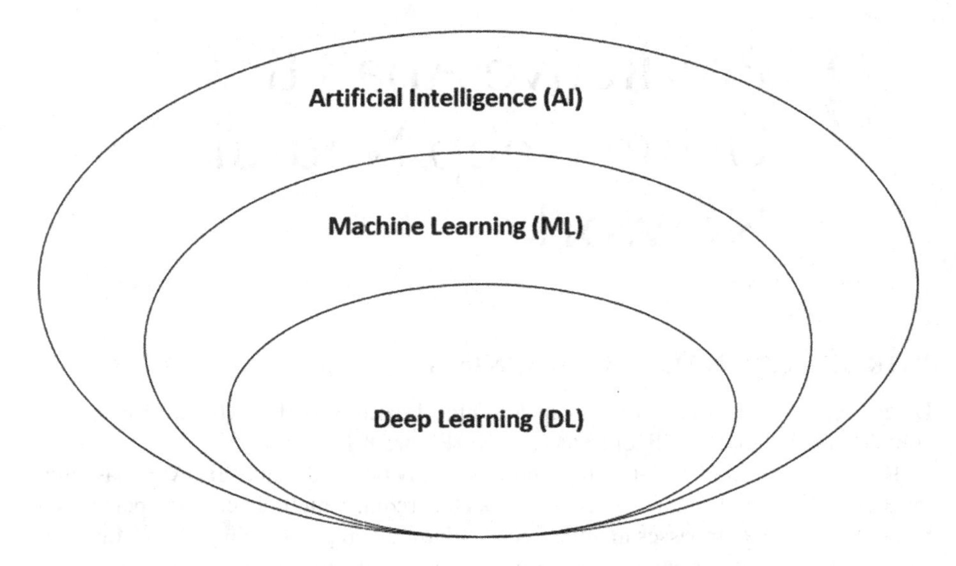

FIGURE 8.1 The relationship between AI, ML, and DL. (Source: Figure adapted from Moolayil, J. (2018). "Learn Keras for Deep Neural Networks: A Fast-Track Approach to Modern Deep Learning with Python." 1st Edition, *Apress*, ISBN 978-1484242391.)

needed because it can better fit the data more accurately using fewer parameters when compared to ANNs (Figure 8.2) [1–3, 5, 6].

The trick here is to keep increasing the number of layers but keeping the number of neurons per layer constant. This will enable the classifier to identify patterns efficiently and accurately.

The disadvantage of a DNN, however, is that the models are harder to train, and they are prone to overfitting. One of the challenges in training DNN is how to efficiently learn the weights of the neurons within each layer. Furthermore, DNN models are complex with a huge number of parameters to train [1–3, 5, 6].

For example, let's consider trying to recognize handwritten text within an image using a DNN classifier. Here, the raw data is the pixel value from an image. The first hidden layer of the DNN captures simple shapes in the image such as the lines and curves. The next hidden layer uses the inputs obtained from the first hidden layer and recognizes higher abstractions, such as corners and circles. Here, the second layer does not have to directly learn from the pixels, which are generally noisy and complex. Therefore, a shallow architecture such as an ANN may require far more parameters because each neuron in a single hidden layer would have limited capabilities to learn about the target variable directly from the pixels of the image [1–3, 5, 6].

DNNs can model complex non-linear relationships and are typically Feed-Forward Neural Networks (FFNNs) in which the data flows from the input layer to the output layer, i.e., only in the forward direction. As mentioned before, the extra layers in the DNN enables the composition of features from lower layers to higher

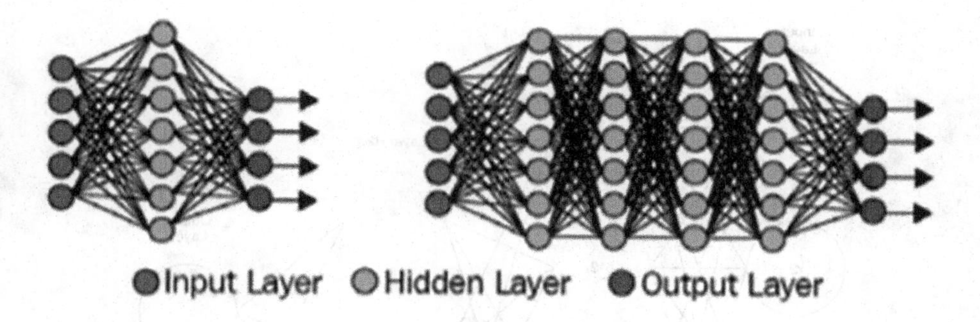

FIGURE 8.2 DNN (left image) and ANN (right image). (Source: Figure adapted from Mostafa, B.M., El-Attar, N., Abd-Elhafeez, S., and Awad, W.A. (2020). "Machine and Deep Learning Approaches in Genome: Review Article," in the *Alfarama Journal of Basic and Applied Sciences*, doi: 10.21608/ajbas.2020.34160.1023.)

layers making it potential for modeling complex data with fewer neurons than a similarly performing ANN network [1–3].

The building block of a DNN is the artificial neuron as shown in Figure 8.3. The concept of artificial neurons was discussed in Chapters 2 and 6. But we will revisit the concept of neurons very briefly for illustrative purposes.

Each of the artificial neuron units receives one or more input signals and outputs a value to the neurons of the following layer and so forth. Then it computes a simple function called the ***activation function***. The activation function sends the processed signal to the next connected neurons. For instance, if the incoming neurons determine a value greater than a threshold (i.e., the activation function is "input > X"), then the output is passed as it is, otherwise it is ignored [1–3].

In a DNN each layer can have one or many neurons. The basic architecture of a 2-layered DNN is shown in Figure 8.4. Here, the first layer is the *input layer* and layer $H1$ and $H2$ are the *hidden layers*. The last layer is called the *output layer* [1–3, 7].

The connection between two neurons of successive layers has an associated **weight**. The weights define the influence of the input on the output for the next neuron and eventually for the overall final output. To begin with the initial weights for each neuron in the DNN model would be a small random number but during the training process, the weights are updated iteratively which enables the DNN model to correctly predict the output.

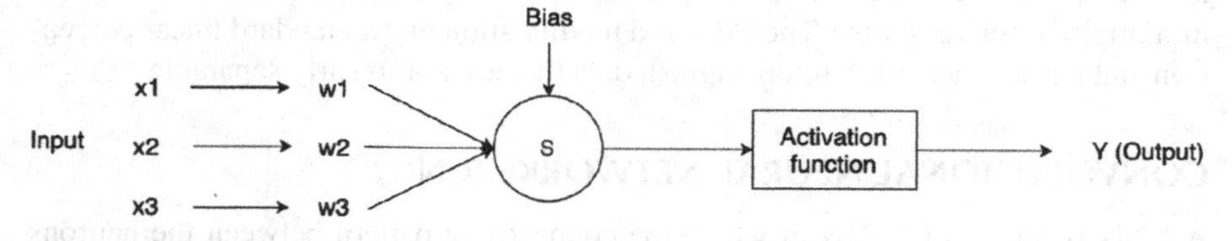

FIGURE 8.3 Artificial neurons as the building blocks of DNN.

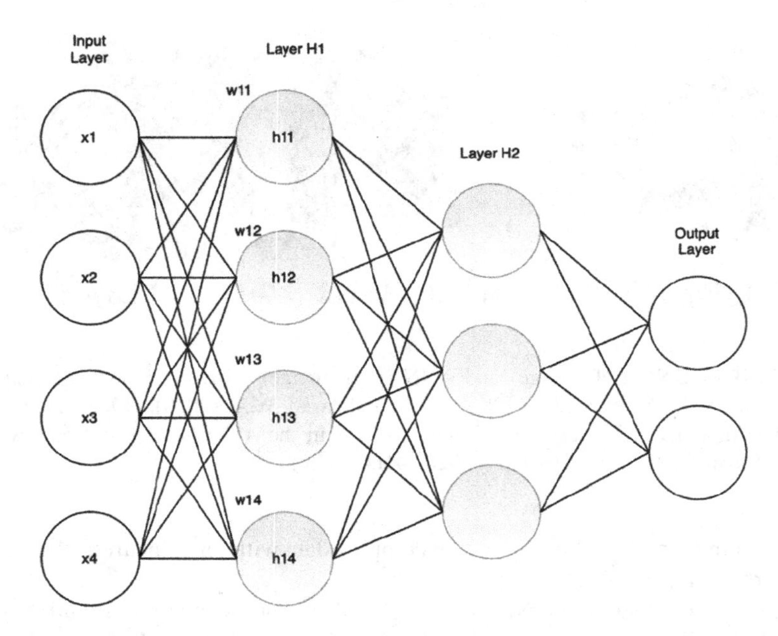

FIGURE 8.4 A typical 2-layer DNN architecture. (Source: Figure adapted from Koutsoukas, A., Monaghan, K. J., Li, X., and Huan, J. (2017). "Deep-learning: investigating deep neural networks hyper-parameters and comparison of performance to shallow methods for modeling bioactivity data," In the *Journal of Cheminformatics*, Vol 9, No. 42, pp. 1–13.)

Now that you have good understanding of the DNN model, let's explore the different architectures of DNN [5, 8, 9].

MULTILAYER PERCEPTRON (MLP)

A *multilayer perceptron (MLP)* is a fully connected, feed-forward ANN model that maps a set of input data to a set of appropriate outputs. An MLP consists of multiple layers of nodes in a directed graph with each layer fully connected to the next layer. Except for the input nodes, each node is a neuron with a nonlinear activation function. The MLP generally utilizes a supervised learning technique, often referred to as *backpropagation* for training the network. In backpropagation a gradient descent function is computed for the feed-forward output and then fed back into the previous layer iteratively. The weights of the previous layer are adjusted at each iteration to minimize the forward error. The MLP is a modification of the standard linear perceptron and has the capability to distinguish data that are not linearly separable.

CONVOLUTIONAL NEURAL NETWORKS (CNN)

A CNN is a type of FFNN in which the connectivity pattern between the neurons within and across the layers are arranged in such a way that they can respond to overlapping regions tiling the visual field. CNNs are inspired by biological processes and are variations of MLP designed to use minimal amounts of preprocessing. CNN's are widely applicable in the area of image and video recognition, recommender systems, and NLP. We will discuss CNNs further in Chapter 9.

RECURRENT NEURAL NETWORKS (RNN)

RNNs are a class of ANNs where connections between the nodes (neurons) form a directed cycle. These directed cycles create an internal state of the network allowing them to exhibit dynamic temporal behavior. Unlike the FFNN, RNN can use their internal memory to process any arbitrary input sequences. Therefore, RNNs find themselves useful for applications such as handwriting and speech recognition. We will discuss more about RNNs in Chapter 10.

AlexNet

AlexNet is one of the earliest and most well-known DNN architecture. In 2012, AlexNet won the difficult ImageNet competition by a large margin. AlexNet scaled the insights of LeNet into a much larger neural network that could be used to learn more complex objects and object hierarchies. The contributions of this architecture include: the use of ReLU as non-linearities, the use of the dropout technique to selectively ignore single neurons during the training process, and to avoid overfitting of the model, overlapping the max pooling layer, avoiding the averaging effects of average pooling, and the use of GPUs NVIDIA GTX 580 to reduce the training time.

VGGNet

VGGNet is a classical CNN architecture with more depth, i.e., with multiple layers to increase the model performance. For example, the VGG-16 or VGG-19 consists of 16 or 19 convolutional layers, respectively. The VGGNet architecture is well-known for its application in object recognition. The insights of this architecture are:

- **Input**: The VGGNet takes an image of input size 224×224.
- **Convolutional Layers**: VGGNet's convolutional layers leverage a minimal receptive field, i.e., 3×3, the smallest possible size that can still capture up/down and left/right motion. Moreover, there are also 1×1 convolution filters acting as a linear transformation of the input. This is followed by a ReLU unit, which is a huge innovation compared to AlexNet which aids in reducing the training time.
- **Hidden Layers**: All the hidden layers in the VGGNet network use the ReLU activation function.
- **Fully Connected Layers**: The VGGNet has three fully connected layers. Out of the three layers, the first two layers have 4096 channels each, and the third layer has 1000 channels, 1 for each class.

INCEPTION

The Inception V3 is an image recognition model that has proven to achieve higher than 78.1 percent accuracy on the ImageNet dataset. This model includes the following symmetric and asymmetric building components which are convolutions, average pooling, max pooling, concatenations, dropouts, and fully linked layers. The batch normalization is performed on the activation inputs and is used extensively throughout the model. Finally, to calculate the loss Softmax function is preferred.

ResNet AND GoogLeNet

In the ResNet model the first two layers are identical to that of the GoogLeNet. The ResNet model includes a 7×7 convolutional layer with 64 output channels, and a stride of 2 which is followed by a 3×3 maximum pooling layer with a stride of 2. After each convolutional layer there is a batch normalization layer. This is the only difference between the ResNet and the GoogLeNet. On the other hand, the GoogLeNet is made up of four modules, each of which is built up of inception blocks. ResNet uses four modules made up of residual blocks, each of which has the same number of output channels. The first module has the same number of channels as the input channel count. It is not essential to reduce the height and width of the residual blocks because a maximum pooling layer with a stride of 2 has already been applied. The number of channels is doubled in the first residual block of each consecutive module relative to the previous module, while the height and breadth are halved. See the Sidebar 2 for more details about the terms related to DNN architecture.

Now let us discuss about the different hyperparameters used in designing the DNN models.

HYPERPARAMETERS OF DNN AND STRATEGIES FOR TUNING THEM

Some of the important hyperparameters to model DNNs are as follows [1, 10]:

ACTIVATION FUNCTION

In DNN, activation function plays an important role in determining the output of the neuron. In most cases, ReLU is preferred as an activation function for the hidden layers. It (ReLU) is a bit faster to compute than any other activation functions. ReLU also ensures that the gradient descent does not get stuck on plateaus as much compared to the logistic function or the hyperbolic tangent function that usually saturates at 1. For the output layer, the softmax activation function is generally a good choice for classification tasks. For regression tasks, no activation functions are preferred. Other commonly used activation functions include Sigmoid and Tanh.

REGULARIZATION

There are several techniques for controlling the training of DNNs in order to prevent overfitting. For example, L2/L1 regularization, max norm constraints, and dropouts. Let's discuss each of the regularization techniques

L2 regularization: This is the most commonly used form of regularization. By updating the gradient descent parameter, the L2 regularization signifies that all weights will eventually decay linearly toward zero.

L1 regularization: In the L1 regularization the absolute value of the weight is penalized. Unlike in L2, the weights may be reduced to *zero* here.

Max-norm constraints: This constraint is issued to enforce an absolute upper boundary on the magnitude of the weight vector for each neuron in the hidden layer. The projected gradient descent is then used to further enforce the constraint.

Dropout: This hyperparameter is tuned while training the model. It is implemented by keeping a neuron active with some probability. For example, the neuron is active when the probability $p < 0.5$, and is zero otherwise. In short, the dropout is a regularization technique which is used to avoid the model to overfit during the training process by randomly dropping few of the nodes in each hidden layer.

NUMBER OF HIDDEN LAYERS

This hyperparameter refers to the number of hidden layers present in the DNN. Upon increasing the number of hidden layers and neurons within each of them, combined with the application of regularization, has a profound effect on the performance of DNN.

NUMBER OF NEURONS PER LAYER

This hyperparameter refers to the number of neurons in each hidden layer of the DNN. Upon increasing the number of hidden layers and neurons within each of them, combined with the application of regularization, has a profound effect on the performance of DNN.

LEARNING RATE

This hyperparameter controls how much change the model should undergo to respond to the estimated error, whenever the model weights are updated. Choosing a value for this hyperparameter is very challenging. If the learning rate is too small it may result in a long training process that could get stuck, whereas a large value for this hyperparameter may result in sub-optimal learning or an unstable training process.

OPTIMIZER

The optimizer is responsible for changing the learning rate and the weights of the neurons in the neural network with an objective to achieve minimal loss. This hyperparameter is very important to achieve the possible highest accuracy or the minimum possible loss.

BATCH SIZE

This hyperparameter is used to speed up the training process. Instead of using the entire dataset for training, this hyperparameter is used for the model to train on the subset of the data. More precisely, the batch size is the number of training data subsamples that is given as an input to the model. The learning process is accelerated by reducing the batch size.

EPOCH

This hyperparameter determines the number of times a dataset is run through the neural network model. In each epoch, the training dataset is transmitted forward and backward through the neural network once. Underfitting can occur when the number of epochs is too small. This is because the neural network has not learned enough. Several to many passes of the training dataset can avoid underfitting. This means that the value for the number of epochs should be high. However, using too many epochs will result in overfitting, where the model can accurately predict the data in the training dataset but not on the test dataset. To achieve the best outcome, the number of epochs must be adjusted or tuned with utmost care.

WEIGHT AND BIASES INITIALIZATION

Initializing the weight and biases of the neurons in the hidden layers is an important hyperparameter to be taken care of. It is recommended not to set all the initial weights to zero. This is because if every neuron in the network computes the same output then there will be no source of asymmetry between the neurons. Therefore, the options would be to initialize the weights of the neurons to a small random number which is not zero. These small numbers can also be drawn from a uniform distribution. In terms of initializing the biases, it is possible and common to initialize the biases to zero since the asymmetry is already taken care of by introducing a small random number for the weights. Setting the biases to a small constant value ensures that all the ReLU units can propagate some gradient.

Now let's look at some strategies for hyperparameter tuning in DNN models.

GRID SEARCH

Grid Search is performed to determine the optimal values of the hyperparameters for a given model. This function helps to loop through the predefined values for the hyperparameters and fit the model on the training dataset. Eventually, this function determines the best values for the listed hyperparameters.

RANDOM SEARCH

The Random Search method replaces the exhaustive search nature of the Grid Search with a random combination of values for the hyperparameters. Since the values are chosen at random, there can be a great deal of variation in the results. However, there is a reduction in the time complexity. Random search is capable of evaluating a large range of values and, in most cases, quickly arrives at a very optimal set of values for the hyperparameters. But the burden of specifying the boundaries for the search area lies on the knowledge worker.

In the next section we will briefly discuss Deep Belief Networks (DBN).

DEEP BELIEF NETWORKS (DBN)

A *Deep Belief Network (DBN)* is a type of DNN with multiple hidden layers and connections between (but not within) layers. This means that a neuron in layer 1 may be connected to a neuron in layer 2 but not with a neuron in layer 1 (Figure 8.5) [1–3, 11].

This restriction of no connections within a layer allows DBN to train much faster using algorithms such as the contrastive divergence algorithm. Essentially, the DBN can be trained layer by layer. To begin with the first hidden layer is trained. During training, the first hidden layer transforms the raw data into a new set of input for the

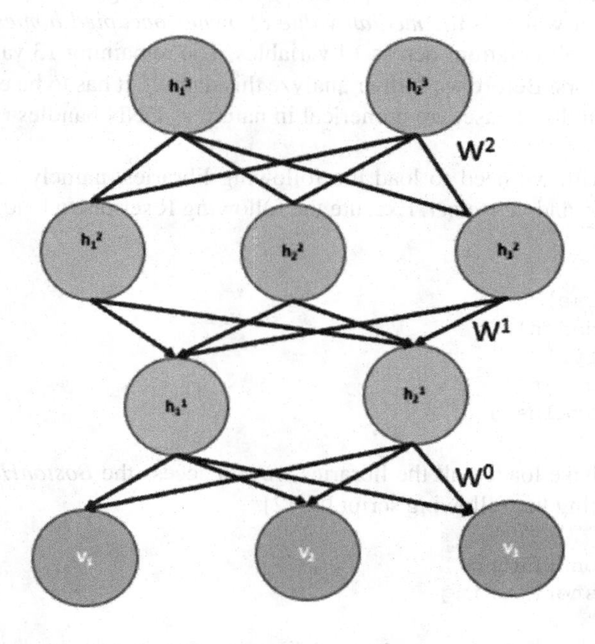

FIGURE 8.5 A typical DBN architecture. (Source: Figure adapted from Kalita, D. (2022). "An Overview of Deep Belief Network (DBN) in Deep Learning," retrieved from https://www.analyticsvidhya.com/blog/2022/03/an-overview-of-deep-belief-network-dbn-in-deep-learning/, retrieved on July 27, 2022.)

next hidden layer. This process is repeated until all the layers have been trained. The benefit of the DBN architecture is that each time a single layer can be trained independent of the other layers. DBNs are sometimes used as a pre-training stage for a DNN [1, 11].

A Restricted Boltzmann Machine (RBM) is a type of generative stochastic ANN that can learn a probability distribution from its inputs. DNNs can also be created using the RBM. DBN, in particular, can be created by "stacking" RBMs and by fine-tuning the resulting deep network via gradient descent and backpropagation. *A series of constrained Boltzmann machines connected in a specific order makes a* **DBN (see Sidebar 3 for more details about RBM)** [1–3, 11].

It's necessary to remember that constructing a DBN necessitates training each RBM layer. To train a complete DBN, the greedy learning technique can be employed. The greedy learning algorithm trains one RBM at a time until all of the RBMs are trained. This section provided a very gentle introduction to the DBN. Extensive discussion of the DBN is beyond the scope of this book [1, 11].

In the next section we will go through a case study that will discuss constructing DNNs and tuning its hyperparameters. The case study will be illustrated using R script.

ANALYZING THE BOSTON HOUSING DATASET USING DNN

In this section, we will design a simple DNN to predict a target variable in the Boston Housing dataset which is the *median value of owner-occupied homes*. This dataset consists of 506 observations across 14 variables. The remaining 13 variables will be used as predictors. Before we further analyze this dataset, it has to be ensured that all the variables in this dataset are numerical in nature as DNN handles only numerical data [12].

To begin with we need to load the following libraries, namely *keras*, *mlbench*, *dplyr*, *magrittr*, and *neuralnet*. Execute the following R scripts to load these libraries [9, 12].

```
library(keras)
library(mlbench)
library(dplyr)
library(magrittr)
library(neuralnet)
```

Now that we have loaded all the libraries we can access the *BostonHousing* dataset in R by executing the following script [9, 12].

```
data("BostonHousing")
data <- BostonHousing
```

Upon executing the R command, the summary of the dataset can be obtained as shown in Figure 8.6 [9, 12].

```
str(data)
```

```
'data.frame':   506 obs. of  14 variables:
 $ crim   : num  0.00632 0.02731 0.02729 0.03237 0.06905 ...
 $ zn     : num  18 0 0 0 0 0 12.5 12.5 12.5 12.5 ...
 $ indus  : num  2.31 7.07 7.07 2.18 2.18 2.18 7.87 7.87 7.87 7.87 ...
 $ chas   : Factor w/ 2 levels "0","1": 1 1 1 1 1 1 1 1 1 1 ...
 $ nox    : num  0.538 0.469 0.469 0.458 0.458 0.458 0.524 0.524 0.524 0.524 ...
 $ rm     : num  6.58 6.42 7.18 7 7.15 ...
 $ age    : num  65.2 78.9 61.1 45.8 54.2 58.7 66.6 96.1 100 85.9 ...
 $ dis    : num  4.09 4.97 4.97 6.06 6.06 ...
 $ rad    : num  1 2 2 3 3 3 5 5 5 5 ...
 $ tax    : num  296 242 242 222 222 222 311 311 311 311 ...
 $ ptratio: num  15.3 17.8 17.8 18.7 18.7 18.7 15.2 15.2 15.2 15.2 ...
 $ b      : num  397 397 393 395 397 ...
 $ lstat  : num  4.98 9.14 4.03 2.94 5.33 ...
 $ medv   : num  24 21.6 34.7 33.4 36.2 28.7 22.9 27.1 16.5 18.9 ...
```

FIGURE 8.6 Summary of the Boston Housing dataset.

The next step is to convert the factor variables into numeric variables using the R command. This function automatically detects all the factor variables and converts them to numerical variables [12].

```
data %<>% mutate_if(is.factor, as.numeric)
```

Now let's create a simple DNN. In this model we will create three hidden layers containing 12, 7, and 5 neurons, respectively.

```
n <- neuralnet(medv ~ crim+zn+indus+chas+nox+rm+age+dis+rad+ta
  x+ptratio+b+lstat,
          data = data,
          hidden = c(12,7,5),
          linear.output = F,
          lifesign = 'full',
          rep=1)
```

Let's view the DNN using the *Plot* command. Note the highlighted area indicating the region containing three hidden layers.

```
plot(n,col.hidden = 'darkgreen',
   col.hidden.synapse = 'darkgreen',
   show.weights = F,
   information = F,
   fill = 'lightblue')
```

The constructed model is shown in Figure 8.7.

Here each predictor variable has one neuron, the first layer has 12 neurons, the second layer has 7 neurons, and the output variable has one neuron.

Next, convert the data frame into matrix for further analysis using the R commands [9, 12]

```
data <- as.matrix(data)
dimnames(data) <- NULL
```

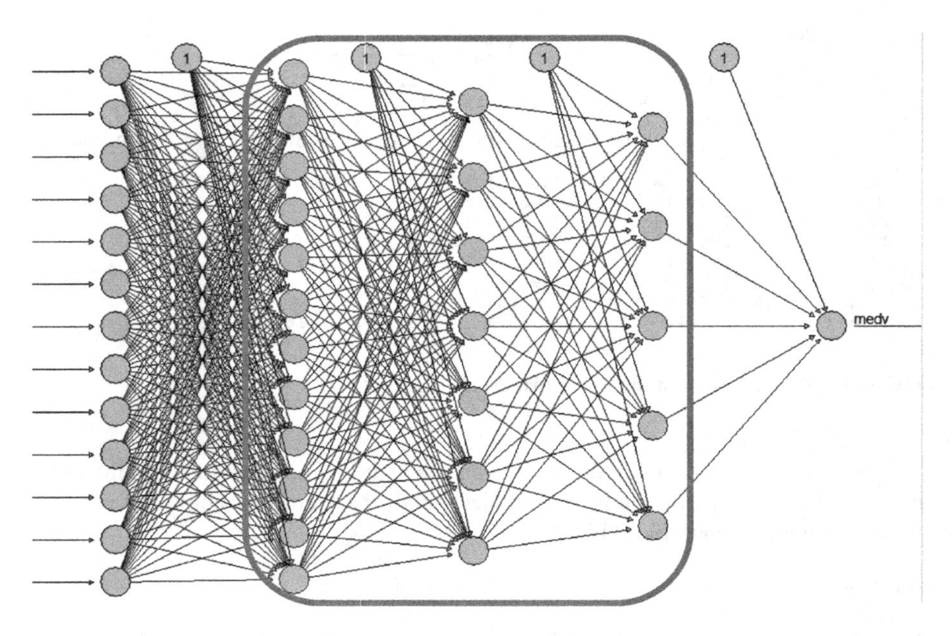

FIGURE 8.7 Simple DNN with three hidden layers consisting of 12, 7, and 5 number of neurons.

Before the DNN model is trained, we need to partition the dataset into training and test dataset. The partition of the dataset is done in the ratio of 70:30, i.e., 70% of the dataset is used for training and the 30% of the dataset is used for testing

```
set.seed(123)
ind <- sample(2, nrow(data), replace = T, prob = c(.7, .3))
training <- data[ind==1,1:13]
test <- data[ind==2, 1:13]
trainingtarget <- data[ind==1, 14]
testtarget <- data[ind==2, 14]
str(trainingtarget)
str(testtarget)
```

Now that the dataset has been portioned, the variables have to be normalized for better prediction. This can be achieved using the following R commands [9, 12]

```
m <- colMeans(training)
s <- apply(training, 2, sd)
training <- scale(training, center = m, scale = s)
test <- scale(test, center = m, scale = s)
```

After completing the preprocessing steps we now look at model building, compiling, and fitting. For building the model, ReLU is used as the activation function, and the hidden layer with 5 neurons is being tested using the 13 predictor variables and 1 output neuron. See the R script below [9, 12].

```
model <- keras_model_sequential()
model %>%
  layer_dense(units = 5, activation = 'relu', input_shape =
  c(13)) %>%
  layer_dense(units = 1)
```

For model compiling, the *rmsprop* is used as an optimizer and the *mae* is used as a metric to measure the performance of the classifier (see the R script below) [9, 12].

```
model %>% compile(loss = 'mse',
      optimizer = 'rmsprop',
      metrics = 'mae')
```

To fit the model the hyperparameters *epochs*, *batch size*, and the *validation split* have been set to 100, 32, and 0.2, respectively, using the following R script [9, 12].

```
mymodel <- model %>%
  fit(training,trainingtarget,
      epochs = 100,
      batch_size = 32,
      validation_split = 0.2)
```

From the graph shown in Figure 8.8, we can see that initially the error is high, but it minimizes toward the end. Over a period of time the output loss is decreasing. After 60 epochs, we note that the testing *mae* is higher than the validation *mae* but the differences are small.

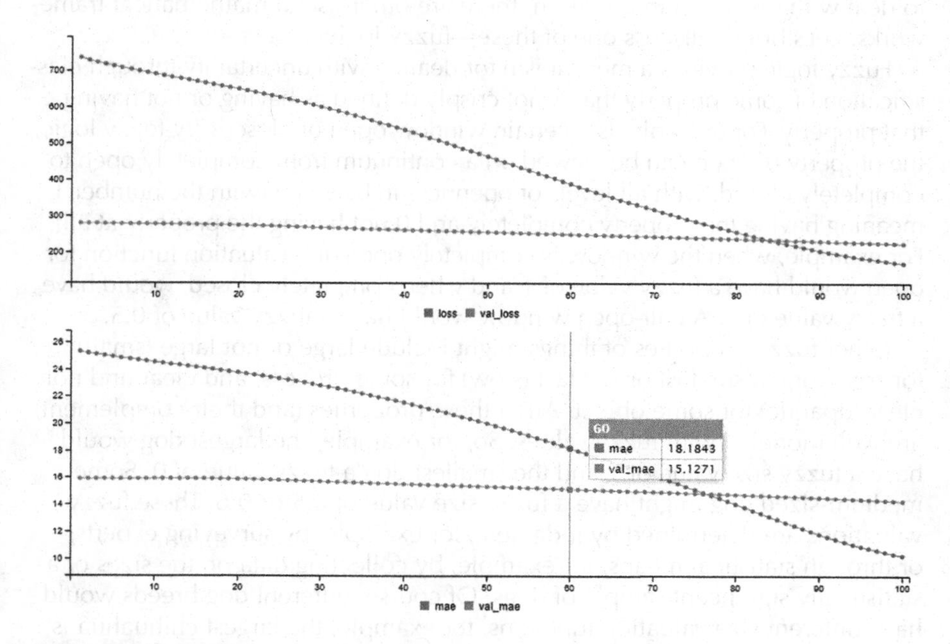

FIGURE 8.8 Testing and validation *mae* across different epochs.

Now let's determine the performance of the prediction on the test dataset by executing the R script [9, 12].

```
model %>% evaluate(test, testtarget)
pred <- model %>% predict(test)
mean((testtarget-pred)^2)
```

Based on the constructed DNN model, the loss (error) is still large around approximately 145. This indicates that by consistently tuning the hyperparameters it is possible to further reduce the error.

SUMMARY

This focus of this chapter was to introduce the concepts of deep learning, the DNN architecture, and the different hyperparameters that can be tuned to effectively perform classification and regression tasks. This is also the first chapter in which we introduce the deep learning concept. Understanding the concepts of this chapter is very important as they lay the basic foundations for the Chapters 9 and 10 where we will discuss the very important Convolutional and Recurrent Neural Networks.

SIDEBAR 1 FUZZY LOGIC AND UNCERTAINTY

Uncertainty has always been a difficult problem for mathematicians, engineers, and scientists. While neural networks and expert systems can be used to deal with uncertain information, there are other useful mathematical frameworks. Let's briefly discuss one of these—fuzzy logic.

Fuzzy logic provides a mechanism for dealing with uncertainty through classification of some property that is not crisply defined as having or not having that property. For example: Is a certain window open or closed? By fuzzy logic, the property of open can be viewed on a continuum from completely open to completely closed, with all levels of openness in between (with the number 1 meaning having the property completely and 0 not having the property at all). For example, when the window is completely open, the valuation function for open would have a fuzzy value of 1 and when completely closed, would have a fuzzy value of 0. A half-open window would have a fuzzy value of 0.5.

Other fuzzy properties of things might include large or not large (small) for some organism, fast or not fast (slow) for some process, and clear and not clear (opaque) for some object. All of these properties (and their complement) are well modeled with fuzzy values. So, for example, the largest dog would have a fuzzy size value of 1 and the smallest dog a fuzzy value of 0. Some medium-sized dog might have a fuzzy size value of 0.5 or 0.6. These fuzzy valuations are determined by judgment, for example, by surveying experts, or through statistical means, for example, by collecting data on the sizes of a statistically significant sample of dogs. Of course, different dog breeds would have different size valuation functions, for example, the largest chihuahua is much smaller than the smallest mastiff.

Boolean operations on fuzzy values need to be redefined as do those on fuzzy sets. For example, for fuzzy values A and B the Boolean AND equivalent is min(A,B) and the OR equivalent is max (A,B). And the Boolean not operation equivalent for A is 1-A.

Fuzzy logic is useful for decision making, in pattern recognition, image processing, and vision problems, and in designing expert systems when the patterns are ill-defined or the input data does not have complete, precise, or reliable information. Many of the data analytics techniques presented in this book can be extended to incorporate fuzzy values.

SIDEBAR 2 TERMS USED IN THE CONTEXT OF DNN ARCHITECTURE

Here we illuminate a few terms related to the DNN architecture, which you will frequently encounter while reading this chapter [1–3, 7, 10].

Activation function: These are the functions that help us to decide if we need to activate the node/neuron or not. These functions are used to introduce non-linearity in the networks.

Convolutional layer: The convolutional layers have a moving filter, which is also referred to as the *weight matrix*. This filter slides over the input image to produce a feature map which is known as the convolution operation. The weight matrix and the input image are multiplied, i.e., the dot product is computed, and summed to produce a feature map. The convolution using another filter over the same image will result in a different feature map.

Dropout: The dropout refers to the process of dropping some neurons during the training phase. Neurons that are dropped are not considered in both the forward and backward pass.

Fully connected layer: In a fully connected layer all the neurons are connected to all the neurons of the previous layer. It is also called the *dense layer*.

Max pooling layer: The window is moved across a 2D input space and the maximum value within that window is considered as the output. It is also called as **down sampling layer** because it reduces the number of parameters within the model.

Padding: Padding is the process of adding extra zeros to the edges of the input matrix. This is done because the contribution of the edge pixels in the input matrix are less than the inside pixels/number.

Strides: The stride parameter is used to decide how the weight matrix should move in the input, i.e., jumping one step or two steps at a time.

Softmax layer: The softmax layer returns the probabilities of each class and the class with higher probability, i.e., the target class. This function calculates the probabilities of each target class over the possibilities of all the target classes.

SIDEBAR 3 RESTRICTED BOLTZMANN MACHINE (RBM)

The *Boltzmann machine* is a network of neurons in which all the neurons are connected to all other neurons. If there are n neurons then connecting each one to every other yields $\frac{n(n-1)}{2}$ connections. In this machine, there are two layers, namely, the visible layer or the input layer and the hidden layer. In the Boltzmann machine, there is no output layer. Boltmann machines are random and generative neural networks capable of learning the internal representations and are able to represent and solve tough combinatoric problems [1–3, 11].

In a Restricted Boltzmann Machine (RBM) the restriction is that the same type of layer is not connected to each other. This means that any two neurons of the input layer or the hidden layer cannot be connected to each other, although the hidden layer and the input layer itself can be connected to each other [1–3, 11].

Remember that the RBM does not have any output layer. Therefore, the question arises, how are we going to identify, adjust the weights, and determine that our prediction is accurate or not? The RBM algorithm learns the probability distribution over its sample training data inputs. The RBM similar to PCA discovers latent factors that can explain the variability in the input data. In short, RBM describes the variability among correlated variables of input dataset in terms of a potentially lower number of unobserved variables. RBMs are widely applicable in the areas of supervised/unsupervised machine learning such as feature learning, dimensionality reduction, classification, collaborative filtering, and topic modeling [1–3, 11].

RBM works in two phases. In the first phase, the input layer is taken and using the concept of weights and biases the hidden layer is activated. This process is termed as the *Feed-Forward Pass (FFP)*. In FFP the positive and the negative association between the visible and hidden unit is determined. Since there is no output layer, in the second phase the attempt is made to reconstruct the input layer through the activated hidden states. This process is termed as *Feed-Backward Pass (FBP)*. Here, in FBP, we are just backtracking the input layer through the activated hidden neurons. After performing FBP the input is reconstructed through the activated hidden state. For this backtracking effort we can then calculate the error and adjust the weights of the neurons in the hidden layer in such a way that the reconstructed input and the original input are close to each other [1–3, 11].

In summary, the DBN is a kind of DNN, which is composed of stacked layers of RBM. It is a random and generative model. Here, the output of each RBM layer acts as the input for the next RBM layer. DBNs can be used to solve unsupervised learning tasks to reduce the dimensionality of features and can also be used to solve supervised learning tasks, i.e., to build classification models or regression models. To train a DBN, there are two steps, layer-by-layer training and fine-tuning. Layer-by-layer training refers to unsupervised training of each RBM, and the fine-tuning refers to the use of error back-propagation algorithms to fine-tune the parameters of the DBN after the unsupervised training is finished [1–3, 11].

EXERCISE

1. The following is TRUE about DNNs
   A. It is easier to train.
   B. The model is simple.
   C. It is prone to overfitting.
   D. The number of neurons per layer cannot vary.
   E. The number of layers is constant in a DNN model.

2. Download the Boston Housing dataset (discussed in lesson 8) and extract a random sample containing 75% of the instances from this dataset. First, summarize this dataset. Second, use the DNN classifier to determine the *median value of owner-occupied homes* using all the 13 predictors. For validation testing, split the dataset into a ratio of 70:30. Clearly highlight the choices for the hyperparameters used in the DNN modeling and summarize the performance of the classifier.

3. Repeat all the instructions provided in question 2. This time vary the number of layers of the DNN and the number of neurons in each layer in an increasing order. Do you notice an improvement over the performance of the DNN classifier with an increase in the number of layers and the number of neurons per layer.

4. Discuss the strategies you will employ to choose an appropriate value for the hyperparameters, batch size, and the number of epochs, while training the DNN.

5. An ___________________ consists of multiple layers of nodes in a directed graph with each layer fully connected to the next layer
   A. Both ResNet and GoogLeNet
   B. Multilayer Perceptron
   C. ResNet
   D. GoogLeNet
   E. None of the above

6. A DNN with multiple hidden layers where the neuron in layer 1 is connected to a neuron in layer 2 but not with a neuron in layer 1 is
   A. DBN
   B. Multilayer Perceptron
   C. CNN
   D. RNN
   E. All the above

REFERENCES

1. Dangeti, P. (2017). *Statistics for Machine Learning*. Packt Publishing Ltd., ISBN 978-1-78829-575-8.
2. Prakash, P. K. S., Achyutuni, S. K. R. (2017). *R Deep Learning Cookbook*. Packt Publishing Ltd., ISBN 978-1-78712-108-9.

3. Hodnett, M., Wiley, JF. (2018). *R Deep Learning Essentials*. 2nd Edition, Packt Publishing Ltd., ISBN 978-1-78899-289-3.
4. Moolayil, J. (2018). *Learn Keras for Deep Neural Networks: A Fast-Track Approach to Modern Deep Learning with Python*. 1st Edition, Apress, ISBN 978-1484242391.
5. Culurciello, E. (2017). "Neural Network Architectures", retrieved from https://towardsdatascience.com/neural-network-architectures-156e5bad51ba, retrieved on July 27, 2022.
6. Mostafa, B. M., El-Attar, N., Abd-Elhafeez, S., Awad, W. A. (2020). Machine and Deep Learning Approaches in Genome: Review Article. *Alfarama Journal of Basic and Applied Sciences*. doi: 10.21608/ajbas.2020.34160.1023.
7. Koutsoukas, A., Monaghan, K. J., Li, X., Huan, J. (2017). Deep-Learning: Investigating Deep Neural Networks Hyper-Parameters and Comparison of Performance to Shallow Methods for Modeling Bioactivity Data. *Journal of cheminformatics*, 9(42), 1–13.
8. Szegedy, C., Vanhoucke, V., Ioffe, S., Shlens, J., Wojna, Z. (2016). Rethinking the Inception Architecture for Computer Vision. *IEEE conference on Computer Vision and Pattern Recognition (CVPR)*, 2818–2826. doi: 10.1109/CVPR.2016.308.
9. LeDell, E. (2018). "user! Machine Learning Tutorial", retrieved from https://koalaverse.github.io/machine-learning-in-R/, retrieved on July 27, 2022.
10. Rendyk. (2021). "Tuning the Hyperparameters and Layers of Neural Network Deep Learning", retrieved from https://www.analyticsvidhya.com/blog/2021/05/tuning-the-hyperparameters-and-layers-of-neural-network-deep-learning/, retrieved on July 27, 2022.
11. Kalita, D. (2022). "An Overview of Deep Belief Network (DBN) in Deep Learning", retrieved from https://www.analyticsvidhya.com/blog/2022/03/an-overview-of-deep-belief-network-dbn-in-deep-learning/, retrieved on July 27, 2022.
12. n.d. (2021). "Deep Neural Network in R", retrievedfrom https://www.r-bloggers.com/2021/04/deep-neural-network-in-r/, retrieved on July 27, 2022.

9 Convolutional Neural Networks (CNN) for Predictive Analytics

We have already discussed the basics of neural networks and deep neural networks. In this chapter, we will discuss the Convolutional Neural Network (CNN). A CNN also referred to as a *ConvNet* is a deep learning algorithm used for classifying images and recognizing objects within images. CNN has provided many successful results both in the area of computer vision and in NLP. For example, Facebook uses CNN as an integral part of their algorithm that automatically tags images, Google uses CNN for searching images, Amazon uses CNN as an integral part of their product recommendation systems, and Instagram uses CNN for image search and recommendations. But CNN can be used for applications beyond just image processing because many problems can be "visualized" as images. For example, the popular genealogical research tool, Ancestry.com, uses CNNs to make inferences about relationships in family trees.

CNNs are inspired by the signal processing capabilities of neurons in the visual cortexes of people and animals. A CNN takes an input image, assigns an importance factor, i.e., learnable weights to various aspects and objects within the image, and then attempts to differentiate one image from another. With most images one can safely assume that all the neighboring pixels are closely related, and that their collective information is more relevant than any individual pixel. This means that each pixel itself does not convey any information about other pixels. For example, in order to recognize letters or digits, we need to analyze the dependency among the neighboring pixels to determine the shape and size of the element. Generally, a pixel in an image is organized in a two-dimensional (2D) grid in *grayscale*. Grayscale means that each pixel can be either black or white (on or off) but the collective effect of a small group of pixels can appear as varying shades of gray. If the image isn't grayscale, it has a third dimension incorporating the pixel color [1, 2].

Now let's consider the motivation behind the use of CNNs for image classification. When feeding an image to a neural network we have to reshape it from 2D to a one-dimensional (1D) array. In this process there can be a significant loss of information. Since we are interested in the information pertaining to those neurons that are closer to the neuron in the question, closely clustered neurons provide more relevant information than the neurons that are further apart [1, 2].

Unlike neural networks, CNNs have the ability to consume information in one, two, or three dimensions and produce an output of the same dimensionality. In addition, CNNs have several benefits when analyzing or classifying images. For example, CNNs connect only those neurons that correspond to the neighboring pixels of the image. Therefore, the neurons take input from only those neurons that are spatially

DOI: 10.1201/9781003278177-9

closed. This has a significant affect in reducing the number of weights since all neurons are not interconnected; whereas in a neural network every neuron is connected to every other neuron in the hidden layer. This aspect results in network not being able to take advantage of the spatial proximity of the pixels in the image. In this case there is no way for the neural networks to determine which pixels are close to each other [1, 2].

Another salient feature of the CNN is parameter sharing, i.e., a limited number of weights are shared among all neurons in a layer. This characteristic further reduces the number of weights and helps fight the overfitting issue. Also, whenever the input contains the same information, then the weights are shared and are neurons are trained jointly.

> In summary one of the main disadvantages of fully connected neural networks (FNN) is that it ignores the structure of the input data. All data feed to the network must be first converted into a 1D numerical array. However, for higher-dimensional arrays such as an image, it gets difficult to deal with such conversion. Therefore, it is essential to preserve the structure of images, as there is substantial hidden information stored inside them. This is where a CNN excels as it considers the original structure of the images while processing them [1, 2].

A CNN's primary function is to compress an image into a format that is easier to process while preserving elements that are important for obtaining a good prediction. The convolutional layer is the most important component of CNN. *Convolution* is an important mathematical operation on two functions that produces a third function, which expresses how the shape of one is modified by the other. In simple words two matrices are multiplied to provide an output that is used to extract features from the image. For example, the convolution between an image (let's say function f) with a filter function, g, will produce a new version of the image. See Sidebar 3 for a more formal description of convolution [1, 2, 3].

CNNs use filters to identify image features, such as edges, straight lines, or arcs. They can also search for certain important patterns or features in an image, for example, a corner. CNNs also exhibit superior performance with image, speech and audio inputs when compared to neural networks.

CNNs are composed of multiple filters or layers containing neurons which perform mathematical functions, i.e., calculate the weighted sum of multiple inputs, and output an activation value of the resultant value. Filters of CNNs are designed to search for certain characteristics in an image and detect whether or not the image contains those characteristics. A filter is applied at different positions in an image, until it covers the entire image. Thus, filters form a very critical element in a convolution layer, which itself is an important operator in a CNN.

There are mainly four different types of layers in a CNN namely [1, 2].

CONVOLUTION LAYER

The convolution layer is the first stage in a CNN. This is the layer where the convolution operation takes place between the input image and the filters. Convolution is carried out to reduce the overall size of the image, so that it is easier to process the image in the following layers.

In order to classify an image, it is important to first identify all the important features. For example, to identify a particular animal it is important to recognize the features, such as eyes, ears, tail, etc. This recognition is what is done in a convolution layer. In this layer, the filter is moved across the image to detect the essential features and the rest are all ignored. Each stage of the image moving through the filters is called a *stride* and the process of moving the image is called *striding* (see Figure 9.1) [1, 2].

The results of the convolution layer are then passed through to the next stage, which is mostly consumed by a non-linear activation function, such as the ReLU function. Figure 9.2 shows an example of a convolution operation on an image [1, 2].

Now let's discuss the parameters used for controlling the convolution operation.

PADDING AND STRIDES

Two parameters, i.e., padding and the strides are used to control the movement of the convolution operation. When a convolution of size $C_1 \times C_2$ is applied to data of size $n \times m$ the output will be $(n - C_1 + 1) \times (m - C_2 + 1)$. If we want the output to be of the same size as the input, then we can pad the input by adding zeros to the borders of the images. This is how the first 3×3 convolution would be applied to the image with padding (see Figure 9.3) [1, 2].

The second parameter that is applied to convolutions are the strides, which control the movement of the convolution. The default value is 1, which means the convolution moves by one step each time, first to the right starting from the left, and then down.

ReLU LAYER

In order to apply non-linearity to the convoluted layer the ReLU function is used, which results in the creation of the convoluted feature map. We learned about the ReLU function in Chapter 2. Images generally have patterns with some level of non-linearity. When convolution is applied on the image there is a risk of losing the non-linear patterns. This is because the convolution operations of multiplication and addition are linear operators. So, a non-linear activation function, such as ReLU, is used to preserve the non-linearity in the images [1, 2].

POOLING LAYER

The pooling layer is used to further reduce the size of the feature representation by applying a function called a *pooling function*. There are different kinds of pooling functions, such as average, max, min etc. Max pooling is widely used as it tends to keep the maximum values of a feature map for each stride [1, 2].

The pooling layer is similar to the convolution layer where a sliding window is used and the window slides over the feature map to find the max value within each stride (see Figure 9.4).

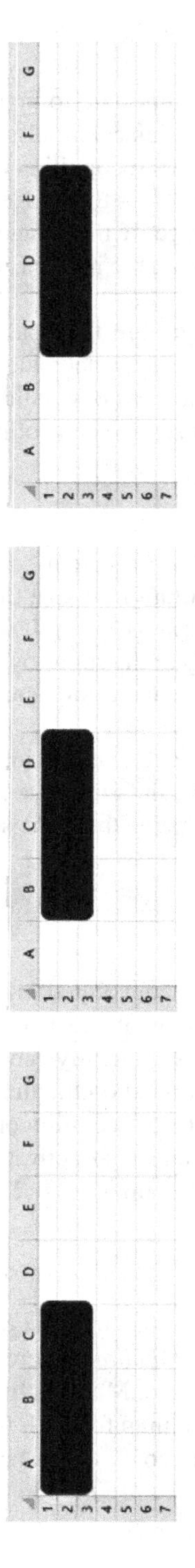

FIGURE 9.1 Striding of the filter over the image one step at a time. (Source: Figure adapted from Hodnett, M., and Wiley, J.F. (2018). "R Deep Learning Essentials." *Packt Publishing Ltd.*, ISBN 978-1-78899-289-3.)

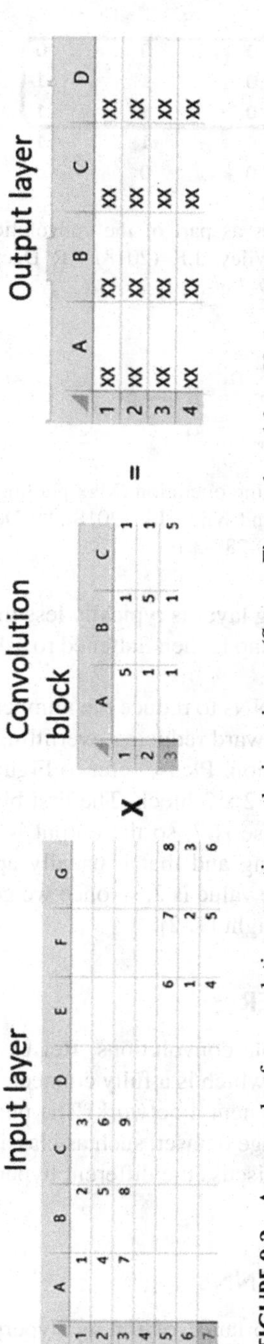

FIGURE 9.2 An example of a convolution operation on the input image. (Source: Figure adapted from Hodnett, M., and Wiley, J.F. (2018). "R Deep Learning Essentials". *Packt Publishing Ltd.*, ISBN 978-1-78899-289-3.)

	A	B	C	D
1	0	0	0	0
2	0	1	1	0
3	0	1	1	0
4	0	1	1	0
5	0	0	0	0

FIGURE 9.3 Padding and strides as part of the convolution operation. (Source: Figure adapted from Hodnett, M., and Wiley, J.F. (2018). "R Deep Learning Essentials." *Packt Publishing Ltd.*, ISBN 978-1-78899-289-3.)

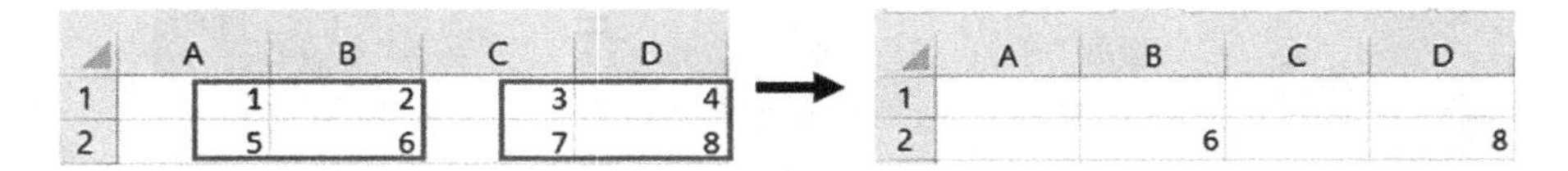

	A	B	C	D
1	1	2	3	4
2	5	6	7	8

	A	B	C	D
1				
2		6		8

FIGURE 9.4 Application of pooling operation (Max pooling) on the input image. (Source: Figure adapted from Hodnett, M., and Wiley, J.F. (2018). "R Deep Learning Essentials." *Packt Publishing Ltd.*, ISBN 978-1-78899-289-3.)

The window size in a pooling layer is typically less than that used in the convolution layer. The pooled feature map is then flattened to a 1D representation to be used in a fully connected layer [1, 2].

Pooling layers are used in CNNs to reduce the number of parameters in the model and therefore they contribute toward reducing overfitting. They can be thought of as a type of dimensionality reduction. Please refer to Figure 9.4 for the following discussion of max pooling using a 2 x 2 block. The first block has the values 7, 0, 6, 6 and the maximum value of these is 7, so the output is 7. Note that padding is not normally used with max-pooling and that it usually applies a stride parameter to move the block. Here, the stride value is 2, so once we get the max of the first block, we move across, 2 cells to the right [1, 2].

FULLY CONNECTED LAYER

In a CNN there can be multiple convolutions, ReLU, and pooling operations but there is only a single last stage which is a fully connected layer. The fully connected layer is the feed-forward of the neural network. The purpose of this step is to make different predictions on the image dataset, such as classifying images [1, 2].

In the next section we will discuss the different hyperparameters that can be used to tune the CNN model.

HYPERPARAMETERS OF CNNS

Let's first understand the importance of tuning hyperparameters in a CNN. In a CNN there can be many convolutional layers which are responsible for sequentially extracting *feature maps* using the convolution kernels. Each convolutional layer can contain several kernels which can convolve the input data from the previous layer to generate the exact amount of scale invariant feature maps. The CNNs have the capability of extracting abstract information that are reserved deeply in raw image (data). Therefore, the number of convolutional layers used in CNN is very much relevant to its feature extraction capability.

The use of the number of kernels in each convolution layer and its size are very relevant to the detail of the information that can be extracted from the image. Application of the non-linear activation function transforms the extracted feature maps elementwise potentially enabling the extraction of more complex features. Therefore, the type of activation function used in each convolutional layer is very much relevant to the convergence speed and to the gradient maintenance during the learning process [4]. These arguments suggest that it is very important to carefully tune the hyperparameters of the CNN model before training it. Table 9.1 lists the key hyperparameters that can be tuned to build efficient CNN models [1, 2, 4].

TABLE 9.1
List of Key Hyperparameters to Tune CNN Models

Hyperparameters	Description	Default Value	Data Type
Number of convolutional layers	Used to specify the number of convolution layers to be used in the CNN model	2	INT
Number of kernels in each convolutional layer	Used to specify the number of kernels in each convolution layer	64	INT
Kernel size in each convolutional layer	Determines the length of the convolution window	3	INT
Activation function in each convolutional layer	Determines whether a neuron should be activated or not by calculating the weighted sum. It introduces non-linearity to the output of a neuron.	ReLU	STRING
Pooling size (if any) after each convolutional layer	Specifies the pooling size of the polling layer after each convolution layer.	2	INT
Number of dense layers	The number of layers between the input layer and the output layer. Usually good to add more layers until the error no longer improves.	2	INT
Connectivity pattern of each dense layer	Specifies the connectivity pattern for the fully connected layer.	Forward	STRING
Number of neurons in each dense layer	Specifies the number of neurons in the fully connected layer.	128	INT
Weight regularization in each dense layer	Specifies the connectivity pattern of each dense layer which involves either L1 (lasso) or L2 (ridge regression) norms that can be introduced to the loss function to regularize each layer's weights to prevent overfitting.	L1	STRING
Dropout rate	Drops out some units of the neural network according to the desired probability to address overfitting (see Sidebar 1)	0.5	FLOAT

(Continued)

TABLE 9.1 (CONTINUED)

List of Key Hyperparameters to Tune CNN Models

Hyperparameters	Description	Default Value	Data Type
Batch size	Defines the number of samples that will be propagated through the network. The weight is updated after every propagation.	50	INT
Epochs	Defines the number of times the training set passes through the neural network, or the number of times the entire training data is shown to the network while training.	10	INT
Learning rate	Specifies how much to change the model in response to the estimated error each time the model weights are updated or how quickly a network updates its parameters.	0.002	FLOAT

In addition to the hyperparameters listed in Table 9.1 there are several general purpose hyperparameters that are used to build CNN models. These additional hyperparameters are [1, 2, 4]:

1. **Learning rules**—these include stochastic gradient descent (SGD), *adagrad*, *adadelta*, *rmsprop*, etc. These rules use the value of a loss function computed over the output of the network, and the expected output, to modify the weights of the neural network in a certain direction (gradient). This effort is performed to reduce the value of the loss function.
2. **Number of filters**—records the depth of the output of convolution, i.e., the dimensionality of the output space in the convolution layer.
3. **Stride**—is the distance on the input matrix that the kernel moves over. A value greater than two is rare. This is an important parameter used in the convolution layer.
4. **Zero-padding** used in the convolution layer. When the filters do not fit the input image, then all the elements that fall outside the input matrix are set to zero. There are 3 types of padding:
   a. **Valid padding**—means no padding.
   b. **Same padding**—means the output is of the same size as the input.
   c. **Full padding**—increases the size of the output by adding zeros.

5. **Weight initialization**—is the process of setting the weights of a neural network to a small random value. For a neural network model, this is the starting point for learning.

In order to determine the appropriate values for the hyperparameters the following search patterns can be followed:

1. **Manual search**—is mostly an *ad hoc* search pattern.
2. **Random search**—a search space is defined as a bounded domain of values. From this domain a random set of points are then sampled for search purposes.

3. **Grid search**—a search space is defined as a grid of hyperparameter values and every position in the grid is evaluated.

In the next section we will perform image classification using a CNN model based on the LeNet architecture (see Sidebar 2). This exercise will also highlight the tuning of the hyperparameters discussed in the previous section.

IMAGE CLASSIFICATION USING A CNN MODEL BASED ON LENET ARCHITECTURE

The *MXNet* package will be used here to realize the CNN model. To install and load the *MXNet* package execute the following R code

```
install.packages("mxnet")
library(mxnet)
```

In addition to the *MXNet* package the following packages should also be installed and loaded in R

```
install.packages("ggplot2","reshape2")
library(ggplot2)
library(reshape2)
```

For classification purpose, we will use the MNIST dataset that can be obtained from this link https://apache-mxnet.s3-accelerate.dualstack.amazonaws.com/R/data/mnist_csv.zip. After downloading the zip file from the link provided above, unzip the folder. The downloaded folder will have two files namely *train.csv* and *test.csv*. Now execute the following R script obtained from [2] to read data from the training dataset and perform image classification using a CNN model.

```
dfFMnist <- read.csv("Location in your hard disk\\mnist_csv\\
  train.csv", header=TRUE)
yvars <- dfFMnist$label
dfFMnist$label <- NULL
```

The training dataset contains images of handwritten digits $(0-9)$, and all of the images are of size 28×28. The dataset contains of a total of 785 columns with the first column containing the data label and the remaining 784 columns containing the pixel values. The training dataset is then split into training and test set to get an unbiased estimate of the accuracy of classification. The split is in the ratio of $90:10$ where 90% of the data is used for training the model and 10% of the data is used for testing the model. Earlier it was indicated that each image is represented as row of 784-pixel values. The value of each pixel is in the range 0 to 255 which is linearly transformed into the range of 0 to 1 by dividing each of the pixel values with 255. The input matrix is then transformed to the column major format. The R script is provided below

```
set.seed(42)
train <- sample(nrow(dfFMnist),0.9*nrow(dfFMnist))
test <- setdiff(seq_len(nrow(dfFMnist)),train)
train.y <- yvars[train]
test.y <- yvars[test]
```

```
train <- data.matrix(dfFMnist[train,])
test <- data.matrix(dfFMnist[test,])
rm(dfFMnist,yvars)
train <- t(train / 255.0)
test <- t(test / 255.0)
```

Both the training and the test datasets are reshaped here so that they are compatible with the data format requirement in the MXNet package.

```
train.array <- train
dim(train.array) <- c(28, 28, 1, ncol(train))
test.array <- test
dim(test.array) <- c(28, 28, 1, ncol(test))
rm(train,test)
```

Now let's output the number of instances for each digit (image) to make sure that the dataset is balanced. Use the following R code:

```
table(train.y)
```

The dataset output looks almost balanced (see Figure 9.5).

The following R script is an implementation of the CNN model based on the LeNet architecture. In this CNN model there are two sets of convolutional and pooling layers and then a flat layer, and finally two dense layers with activation and dropout.

```
act_type1="relu"
devices <- mx.cpu()
mx.set.seed(0)
data <- mx.symbol.Variable('data')

# The First convolution layer
convolution1 <- mx.symbol.Convolution(data=data,
  kernel=c(5,5),
num_filter=64)
activation1 <- mx.symbol.Activation(data=convolution1,
  act_type="tanh")
pool1 <- mx.symbol.Pooling(data=activation1, pool_type="max",
kernel=c(2,2), stride=c(2,2))

# The Second convolution layer
convolution2 <- mx.symbol.Convolution(data=pool1,
  kernel=c(5,5),
num_filter=32)
activation2 <- mx.symbol.Activation(data=convolution2,
  act_type="relu")
pool2 <- mx.symbol.Pooling(data=activation2, pool_type="max",
kernel=c(2,2), stride=c(2,2))
```

```
train.y
   0    1    2    3    4    5    6    7    8    9
3716 4229 3736 3914 3672 3413 3700 3998 3640 3782
```

FIGURE 9.5 Number of instances for each digit (image) in the training dataset.

```
# The flatten layer and then fully connected layers
flatten <- mx.symbol.Flatten(data=pool2)
fullconnect1 <- mx.symbol.FullyConnected(data=flatten,
  num_hidden=512)
activation3 <- mx.symbol.Activation(data=fullconnect1,
  act_type="relu")
fullconnect2 <- mx.symbol.FullyConnected(data=activation3,
  num_hidden=10)
# final softmax layer
softmax <- mx.symbol.SoftmaxOutput(data=fullconnect2)
```

Please refer to Sidebar 2 for more discussions on the LeNet and other architectural options for building CNN models.

Now let's run the model. The R script for running the model is as follows:

```
devices <- mx.gpu()
mx.set.seed(0)
model2 <- mx.model.FeedForward.create(softmax, X=train.array,
  y=train.y,
ctx=devices,array.batch.size=128,
num.round=10,
learning.rate=0.05, momentum=0.9,
wd=0.00001,
eval.metric=mx.metric.accuracy,
epoch.end.callback=mx.callback.log.train.metric(1))
```

Now we'll evaluate the model by obtaining the confusion matrix and the performance measure, accuracy, which are both shown in Figures 9.6 and 9.7, respectively.

The R script for evaluating the performance of the CNN model based on the LeNet architecture is as follows:

```
# evaluate model
preds2 <- predict(model2, test.array)
pred.label2 <- max.col(t(preds2)) - 1
res2 <- data.frame(cbind(test.y,pred.label2))
table(res2)
accuracy2 <- sum(res2$test.y == res2$pred.label2) / nrow(res2)
accuracy2
```

We note here that the classification accuracy is around 98%.

Let's quickly examine the details of the CNN model built in this exercise. The architecture of LeNet has been programmed using the MXNet package. Here, the LeNet architecture has two convolutional groups and two fully connected layers.

```
pred.label2
test.y   0   1   2   3   4   5   6   7   8   9
     0 412   0   0   0   0   1   1   1   0   1
     1   0 447   1   1   1   0   0   4   1   0
     2   0   0 438   0   0   0   0   3   0   0
     3   0   0   6 427   0   1   0   1   2   0
     4   0   0   0   0 395   0   0   1   0   4
     5   1   0   0   5   0 369   2   0   1   4
     6   2   0   0   0   1   1 432   0   1   0
     7   0   0   2   0   0   0   0 399   0   2
     8   1   0   1   0   1   1   1   1 414   3
     9   2   0   0   0   4   0   0   1   1 398
```

FIGURE 9.6 Confusion matrix demonstrating the performance of the CNN model based on the LeNet architecture.

```
accuracy'
   0.9835714
```

FIGURE 9.7 Classification accuracy of the CNN model based on the LeNet architecture.

The convolutional groups in-turn have a convolutional layer, followed by an activation function and then a pooling layer. This combination of layers is very common for image classification tasks. The first convolution layer has 64 blocks of 5×5 size with no padding. The pooling layers have been configured with the Max pooling capability. The stride is configured to 2. The fully connected layer has two layers, one with 512 nodes and the other with 10 nodes. Finally, the *softmax* activation function is used to convert the numeric quantities in this layer into a set of probabilities for each category [1, 2].

SUMMARY

This focus of this chapter was to introduce to the CNN architecture and the different hyperparameters which can be tuned to effectively perform image classification. To this end only the LeNet architecture was discussed here in detail in addition to explaining the general components of the CNN architecture. CNNs are best suited for the image classification but recent advances in the CNN architecture designs including the GoogLeNet and ResNet have found many applications in the area of NLP.

SIDEBAR 1 DROPOUT

It is a form of regularization which aims at preventing the model from overfitting. Overfitting occurs when the model attempts to memorize parts of the training dataset but is not as accurate on unseen test data. Overfitting can be checked by determining the gap between the accuracy on the training set against the accuracy on the test set. If performance is much better on the training dataset compared to the test dataset, then the model is overfitting. Dropout refers to removing nodes randomly (based on random distribution) from a network temporarily during the training process. It is usually only applied to hidden layers, and not on the input layers. During each forward pass, a different set of nodes is removed, thus the network is different each time. Another way to look at dropout is that each node in a layer must learn to work with all the nodes in that layer. Dropouts also result in preventing one or more number of nodes in a layer from getting large weights and dominating the outputs from that layer [1, 2].

SIDEBAR 2 CNN ARCHITECTURES [1, 2, 3]

There are four different types of CNN architectures namely LeNet, AlexNet, GoogLeNet, and ResNet. Here, we will briefly discuss about each of the above-mentioned architectures.

LeNet: In the LeNet architecture the network has 5 layers with learnable parameters and hence is named as LeNet. LeNet consists of an input layer, 3 sets of convolutional layers with 2 average pooling layers in between them. After the convolutional and pooling layer, there are 2 fully connected layers. One of the two is an output layer.

AlexNet: The AlexNet architecture is very similar to the LeNet architecture. However, AlexNet is much larger and deeper, and it stacks convolutional layers directly on top of each other instead of stacking a pooling layer on top of each convolutional layer. AlexNet has 8 layers with learnable parameters. The architecture of AlexNet consists of an input layer, 2 sets of combination of convolution and max pooling layers, 3 convolutional layers, a max pool layer, 2 sets of combination of dropout and fully connected layer followed by an output fully connected layer.

GoogLeNet: GoogLeNet is the CNN architecture used by Google to win ILSVRC (ImageNet Large Scale Visual Recognition Challenge) 2014 classification task. The architecture consists of a 22-layer deep CNN with small convolutions, called "inceptions," batch normalization, and other techniques to reduce the number of parameters from 60 million in previous architectures to four million. GoogLeNet CNN architecture is computationally expensive. To reduce the parameters that must be learned, it uses heavy unpooling layers on top of CNNs to remove spatial redundancy during training and also incorporates shortcut connections between the first two convolutional layers before adding new filters in later CNN layers.

ResNet: ResNet is a CNN architecture that was developed by Kaiming et al. to win the ILSVRC 2015 classification task. There are 152 layers in the network, with over one million parameters. Like GoogLeNet, it uses a lot of batch normalization. ResNet employs an innovative design which allows it to run many more convolutional layers without increasing the complexity. CNNs are mostly used for image classification tasks with 1000 classes, but ResNet shows that CNNs can also be used successfully to solve NLP problems like sentence completion or machine comprehension, where it was used by the Microsoft Research Asia team in 2016 and 2017, respectively. The CNN architecture ResNet is computationally efficient and can be scaled up or down to match computational power of GPUs.

SIDEBAR 3 CNN AND VOLTERRA SERIES

The convolution operation is one of the most important tools in signal and image processing and many other engineering applications. It has important theoretical properties and practical applications. Let's formalize the notion of convolution and relate it to an important multidimensional system concept that has important applications, including in data analytics [5, 6].

Let u be a discrete time input function that is bounded and time limited. Also consider a system represented by h, which is another discrete time, bounded and time-limited function. While h, which is often called a *kernel*, could represent any kind of system, it could be some kind of filter. In any case, the one-dimensional *discrete convolution* of u with h, denoted $(h*u)(k)$ is given by an infinite sum [5, 6]:

$$y(k) = (h*u)(k) = \sum_{j=-\infty}^{\infty} h(j)u(k-j)$$

where the output y is called the response of the system (or kernel) h to input u. This is the standard one-dimensional convolution form. There are also continuous versions of this convolution, but we are interested in the discrete form only.

Higher-order convolutions can be obtained as well. Consider, for example, the discrete two-dimensional input function $u(k_1,k_2)$ and the discrete two-dimensional kernel h. Then the two-dimensional convolution of u with h, is denoted

$$y(k_{1,},k_2) = (h*u)(k_{1,},k_2) = \sum_{j_1=-\infty}^{\infty}\sum_{j_2=-\infty}^{\infty} h(j_1,j_2)u(k_1-j_1,k_2-j_2)$$

For our purposes, both u and h are bounded and time-limited functions. This form is convenient for two-dimensional digital images. In a similar way 3-, 4-, and higher-order convolutions can be expressed [5, 6].

Now consider a system described by an infinite sum of homogeneous terms and multidimensional convolution operations:

$$y(k) = \sum_{n=1}^{N}\sum_{j_n=-\infty}^{\infty}\cdots\sum_{j_1=-\infty}^{\infty} h_n(j_1,\ldots,j_n)u(k-j_1)\ldots u(k-j_n)$$

is called a discrete *Volterra system of order N*. Here, $y(k)$ is also known as a *Volterra series*. The h_i are called *kernel functions* and the $u()$ form a class of degree-n input functions. To ensure convergence, assume all functions are bounded and have compact support. There is also a continuous formulation of Volterra systems [5, 6].

The Volterra series was discovered in the early 1880s but were only a mathematical curiosity until the 1940s when Norbert Weiner began to use them to model control systems. Since then, they have been used in signal and image processing and systems analysis. One of this text's authors was studying the relationship of Volterra series to morphological (set based) image processing and massive parallel computers in the late 1980s as part of his doctoral research [5, 6]. This work largely lay dormant due to the previously noted AI winter. More recently, however, Volterra series have been used to model perceptrons [Marmarelis, Vasilis Z., and Xiao Zhao. "Volterra models and three-layer perceptrons." *IEEE Transactions on Neural Networks* 8.6 (1997): 1421–1433]. Li et al. recently showed that convolutional neural networks can be approximated by a finite Volterra series, whose order increases exponentially with the number of layers and kernel size increases exponentially. [Li, Tenghui, et al. "Understanding Convolutional Neural Networks from Theoretical Perspective via Volterra Convolution." *arXiv preprint arXiv:2110.09902* (2021).]

Perhaps Volterra series will play a further role in data analytics and AI, clearly showing that some theoretical concepts can take centuries or more to find real-worsld applications.

EXERCISE

1. Unlike neural networks, ___________ has the ability to consume information in one-,two-, or three-dimensions and produce an output of the same dimensionality
   A. ANN
   B. RNN
   C. DNN
   D. CNN
   E. CNN and RNN

2. Download the MNIST dataset (discussed in Chapter 9). First, summarize this dataset. Second, use the CNN model based on the AlexNet architecture to determine the label of the images using all the predictors. For validation testing, split the dataset into a ratio of 80:20. Clearly highlight the choices for the hyperparameters used in the AlexNet-based CNN modeling and summarize the performance of the classifier.

3. Repeat all the instructions provided in question 2. This time build the CNN model based on the ResNet architecture to determine the label of the images

on the MNIST dataset using all the predictors. Compare the performance of the ResNet architecture and the AlexNet architecture–based CNN models.

4. Compare the different architectures of CNN for image processing applications?

5. One of the listed hyperparameters is used to drop units of neural networks according to the desired probability to address overfitting
   A. Epochs
   B. Dropout rate
   C. Learning rate
   D. Batch size
   E. None of the above

6. Discuss the importance of the pooling layer? What are the different types of pooling functions that are available?

REFERENCES

1. Dutta, S. (2018). *Reinforcement Learning with TensorFlow*. Packt Publishing Ltd., ISBN 978-1-78883-572-5.
2. Hodnett, M., Wiley, J. F. (2018). *R Deep Learning Essentials*. Packt Publishing Ltd., ISBN 978-1-78899-289-3.
3. Zhu, W., Yeh, W., Chen, J., Chen, D., Li, A., Lin, Y. (2019). "Evolutionary Convolutional Neural Networks Using ABC," *Proceedings of the 11th International Conference on Machine Learning and Computing*, pp. 156–162.
4. Liu, P., Zeng, Z., Wang, J. (2017). Multistability of Delayed Recurrent Neural Networks with Mexican Hat Activation Functions. *Neural Computation*, 29(2), 423–457.
5. He, K., Zhang, X., Ren, S., Sun, J. (2015). "Deep Residual Learning for Image Recognition". doi:10.48550/arXiv.1512.03385.
6. Laplante, P, (1990). *On Volterra series and morphological operations*. PhD Dissertation, Stevens Institute of Technology, Hoboken, NJ.

10 Recurrent Neural Networks (RNNs) for Predictive Analytics

In earlier chapters we discussed the basics of Deep and Convolutional Neural Networks. Our next frontier will be to explore another well-known deep neural network architecture—*Recurrent Neural Networks*.

RECURRENT NEURAL NETWORKS

A Recurrent Neural Network (RNN) is a type of artificial neural network that uses sequential or time series data as an input, output, or both. RNNs are very effective because of their architecture, which aggregates the learning from the past datasets and uses them along with the new data to enhance the learning process [1–4]. This unique functionality helps them to capture the sequence of events, which wasn't possible in other (feed forward) neural networks. For example, consider a time series data related to audio, video, or the stock market where the sequence of events matters a lot. Collective learning from these datasets can help the model to capture the underlying trend. The ability to perform sequence-based learning is what makes RNNs highly effective.

Let's discuss a problem related to sequence-based learning and see how RNNs can provide the solution to the problem. Imagine a sequence of events at each point in time on which we need to decide about the sequence of events. If the sequence is stationary, then a classifier with similar weights for any time step can be enough to predict the outcome. However, if the same classifier is used at different time step, then the weights of the classifier will differ [4].

Conversely, when training is performed on the entire dataset containing data for all time steps then the weights of the classifier will be the same, but the sequence-based learning will be hampered. Therefore, our solution should be to share the weights over different time steps and also use what we have learned in previous time steps up to and including the last step.

It is well-known that neural networks should be able to consider learning from the past. In order to remember further back, a deeper neural network would be required. This neural network would need a single model that can summarize the past and provide that information, along with the new information, to the classifier. Therefore, in an RNN with input vector X and output vector Y, the following parameters are computed at any time step t:

$$h_t = \tanh\left(W_h\left[h_{t-1}; X_t\right] + b_h\right)$$

DOI: 10.1201/9781003278177-10

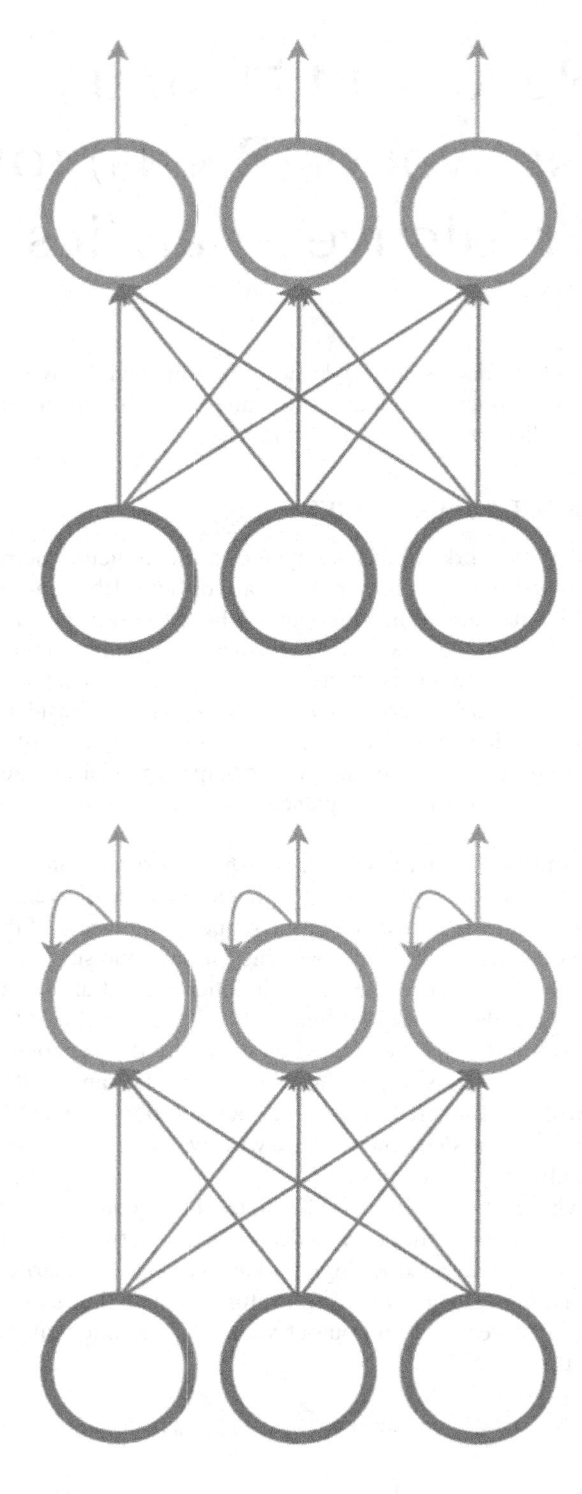

FIGURE 10.1 Comparison of RNN (on the left) and feed-forward neural networks (on the right). (Source: Figure adapted from Eliasy, A., Przychodzen, J. (2020). "The role of AI in capital structure to enhance corporate funding strategies." *Array* 6, 100017, ISSN 2590-0056, https://doi.org/10.1016/j.array.2020.100017, 1–13.)

where W_h and b_h are the *weights* and *biases* shared over time, tanh is the activation function, and $h_{t-1};X_t$ refers to the concatenation of the two sets of information. Here, the dimension of the X_t is $n \times d$ where n is the number of samples or rows in the dataset and d is the number of dimensions or columns. The dimension h_{t-1} of is given by $n \times l$ and the concatenation of the two sets of information will result in a matrix of size $n \times (d + l)$ [2].

Once the forward propagation task is completed, the next task would be to minimize the overall loss by backpropagation. The total loss is the summation of loss across all time steps. Therefore, given a sequence of X values and the corresponding sequence of output Y values, the total loss is given as [2]:

$$L = \sum_{i=1}^{t} L_i = \sum_{i=1}^{t} \left[-y \log(\hat{y})\right]_i$$

where $\hat{y}$ is the predicted output.

As noted, RNNs are commonly used for ordinal or temporal problems, such as language translation, Natural Language Processing (NLP), speech recognition, and image captioning. These applications can be found in popular applications such as Apple's Siri intelligent assistant, many voice searching apps, and Google Translate.

It is important to understand the main difference between the RNN and feedforward CNNs. RNNs are distinguished from their counterparts because of their use of memory. RNNs take information from prior inputs to influence the current input and output. In Figure 10.1, we can see that the output of the RNN depends on the prior elements within the sequence. This means that the future events could be helpful in influencing or determining the output of a given sequence. Unidirectional RNNs don't account for these events in their predictions [5].

Another distinguishing characteristic of RNNs is that they share parameters across each layer of the network—feed-forward networks have different weights across each node. However, in the RNN the weights of the parameters are still adjusted in the process of backpropagation and gradient descent to facilitate reinforcement learning.

Now let's discuss one of the most prominent structures of the RNN, which is *Long Short-Term Memory*.

Long Short-Term Memory

Generally, RNNs fail to handle dependencies on a long-term basis. When the distance between the output data point in the output sequence and the input data point in the input sequence increases, the RNNs fail in connecting the relationship or information between them. This phenomenon is common in text-based tasks where the length of sequences is long. The Long Short-Term Memory (LSTM) architecture, shown in Figure 10.2, is capable of handling these long-term dependencies [1, 3, 6].

The LSTM architecture has a chain structure that contains four neural networks and different memory blocks called cells. The main idea of LSTM networks is the cell state, which is the line that runs horizontally through the structure of LSTM in

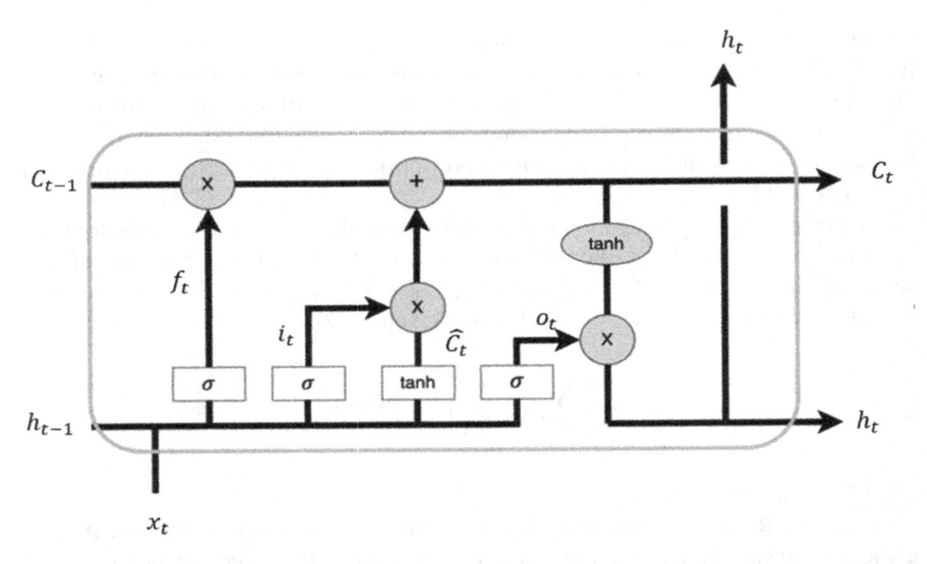

FIGURE 10.2 Internal structure of LSTM. (Source: Figure adapted from https://www.tutorialexample.com/understand-the-effect-of-lstm-input-gate-forget-gate-and-output-gate-lstm-network-tutorial/.)

Figure 10.2. It can be seen that it works as a conveyor belt moving through the entire chain with only limited minor linear interactions, so that the information flows throughout without any change. The other arrows function as doors that interact with the cell status observing what information is really relevant to allow its passage. In a LSTM the information is retained by the cells and the memory manipulations are done by the gates. There are three gates represented as σ (see Figure 10.2) namely:

Forget Gate

The information that is no longer useful in the cell state of a LSTM is removed using the *forget gate*. Figure 10.3 shows the cross-section view of the LSTM of Figure 10.2 highlighting the forget gate.

In Figure 10.3, two inputs x_t (input at a particular time) and h_{t-1} (output from the previous cell) are fed to the gate and multiplied using the weight matrices followed by the addition of bias. The resultant value is passed through an activation function which gives a binary output. Note here that based on the output of the cell state, either the piece of information is forgotten or retained.

Input Gate

The addition of useful information to the cell state of a LSTM is accomplished by an *input gate*. Figure 10.4 shows the cross-section view of the LSTM of Figure 10.2 highlighting the input gate.

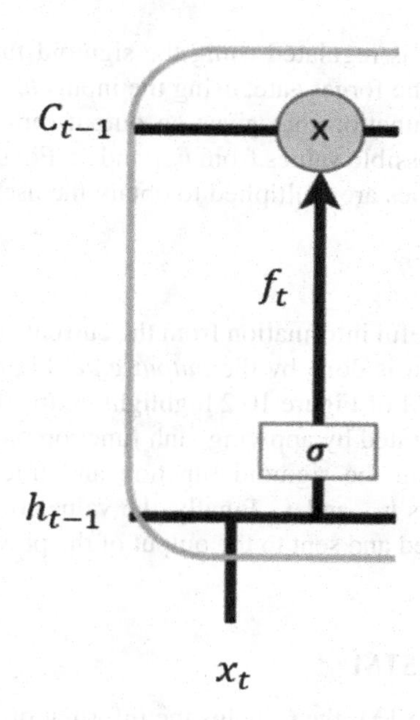

FIGURE 10.3 The forget gate. (Source: Figure adapted from https://www.tutorialexample.com/understand-the-effect-of-lstm-input-gate-forget-gate-and-output-gate-lstm-network-tutorial/.)

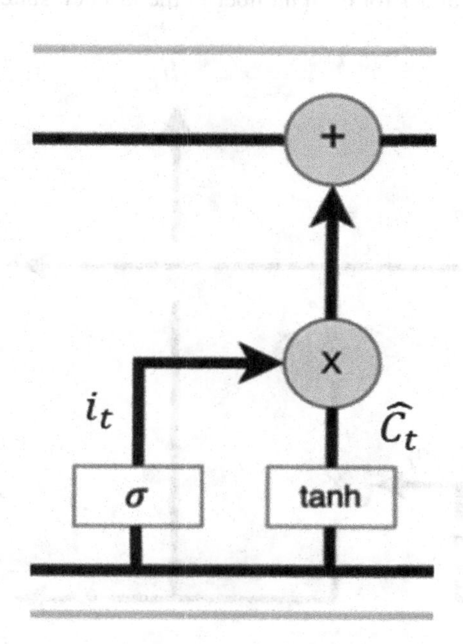

FIGURE 10.4 The input gate. (Source: Figure adapted from https://www.tutorialexample.com/understand-the-effect-of-lstm-input-gate-forget-gate-and-output-gate-lstm-network-tutorial/.)

First, the information is regulated using the sigmoid function and filtered to be remembered, similar to the forget gate, using the inputs h_{t-1} and x_t. Next, a vector is created using the tanh function that gives an output ranging between −1 and +1, which contains all the possible values from h_{t-1} and x_t. Finally, the values of the vector and the regulated values are multiplied to obtain the useful information.

Output Gate

The task of extracting useful information from the current cell state of the LSTM and presenting it as an output is done by the *output gate*. Figure 10.5 shows the cross-section view of the LSTM of Figure 10.2 highlighting the output gate.

First, a vector is generated by applying tanh function on the cell. Then, the information is regulated using the sigmoid function and filtered by the values to be remembered using inputs h_{t-1} and x_t. Finally, the values of the vector and the regulated values are multiplied and sent to the output of the previous cell and as input for the next cell.

More Details of the LSTM

The key feature of the LSTM, the C_t helps the information to flow unchanged. Now let's discuss the forget gate layer, namely f_t. The forget gate concatenates the last hidden layer, i.e., h_{t-1} and the input layer x_t and trains the neural network, resulting in a number between 0 and 1 for each number in the last cell state C_{t-1}, where 1 means

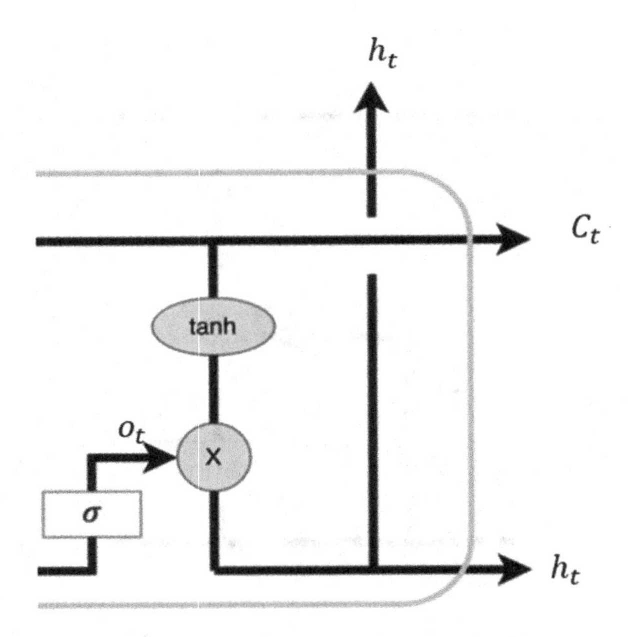

FIGURE 10.5 The output gate. (Source: Figure adapted from https://www.tutorialexample.com/understand-the-effect-of-lstm-input-gate-forget-gate-and-output-gate-lstm-network-tutorial/.)

to keep the value and 0 means to forget the value. Therefore, the forget layer is used to identify the information that needs to be retained and the information that needs to be forgotten [2]. Mathematically,

$$f_t = \sigma\left(W_f\left[h_{t-1}; x_t\right] + b_f\right)$$

Next, let's look at the input gate layer i_t and the tanh layer $\hat{C}_t$, which identify what new information needs to be added into one received from the past to update the information, i.e., the cell state. The tanh layer results in a vector of values while the input gate layer identifies which of those values to use for updating the information. Therefore, the cell state C_t is the combination of the new information discussed above and the information retained by the forget gate layer. Mathematically,

$$i_t = \sigma\left(W_i\left[h_{t-1}; x_t\right] + b_i\right)$$

$$\hat{C}_t = \tanh\left(W_C\left[h_{t-1}; x_t\right] + b_C\right)$$

Therefore,

$$C_t = f_t \times C_{t-1} + i_t \times \hat{C}_t$$

The RNN is trained at the output gate layer o_t which can be represented as [2]

$$o_t = \sigma\left(W_o\left[h_{t-1}; x_t\right] + b_o\right) \text{ and } h_t = o_t \times \tanh\left(C_t\right)$$

In summary, the LSTM cell incorporates the last cell state C_{t-1}, last hidden state h_{t-1}, and the current time step input x_t, and outputs the updated cell state C_t and the current hidden state h_t.

In a way, LSTM is believed to be a modification to the RNN hidden layer, which could overcome the vanishing gradient problem (see Side bar 1) that exists in RNNs (see Figure 10.6) [4].

LSTMs work extremely well in a wide variety of problems. They are explicitly designed to avoid problem of long-term dependency. This is accomplished by remembering information for long periods of time as part of their basic behavior.

Hyperparameters for RNNs

In this section, we will discuss about the different hyperparameters of the RNN classifier. See Table 10.1 for the list of the hyperparameters used in architecting the RNN classifier, their characteristics, and their descriptions [1–3, 6].

Let's consider the task of text classification on the Reuters dataset to illustrate the potential of the RNN (LSTM) classifier. The Reuters dataset can be accessed through a function in the *Keras* library. This dataset consists of 11,228 records with 46 categories.

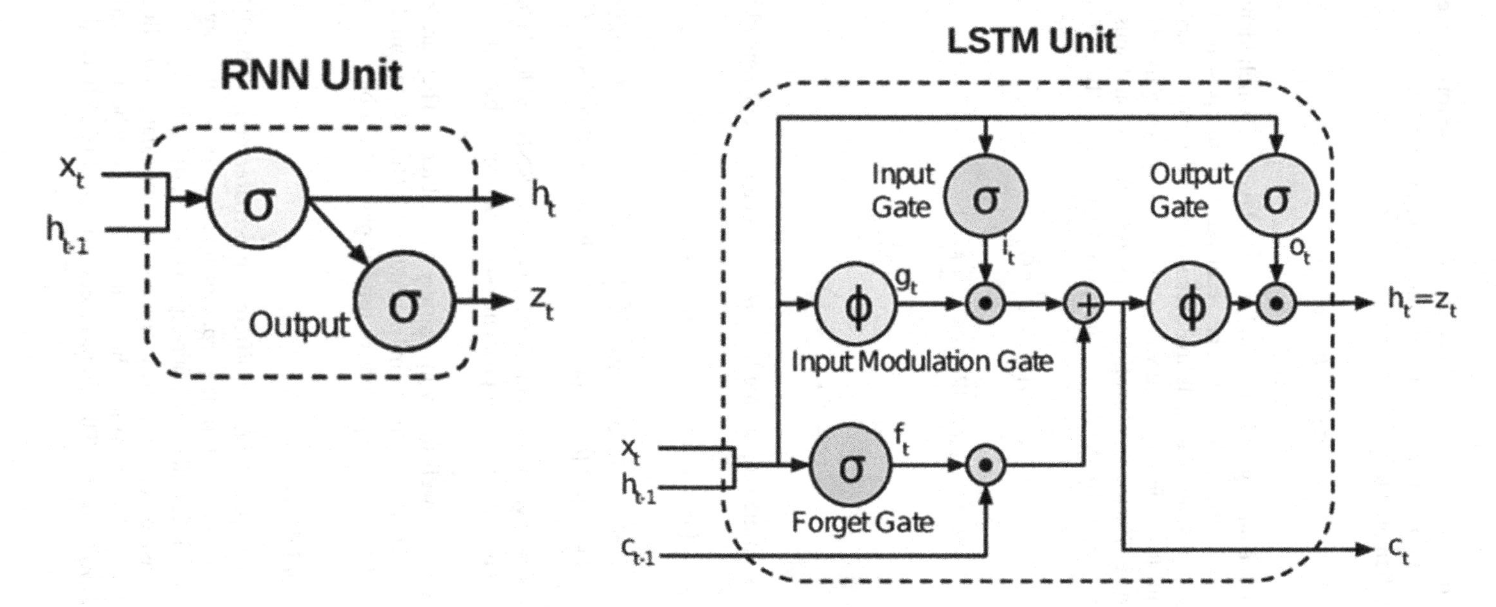

FIGURE 10.6 Internal structure of the basic RNN and LSTM unit. (Source: Figure adapted from Jayawardhana, S. (2020). "Sequence Models & Recurrent Neural Networks (RNNs)," Retrieved from https://towardsdatascience.com/sequence-models-and-recurrent-neural-networks-rnns-62cadeb4f1e1, retrieved on June 18, 2022.)

TABLE 10.1
Hyperparameters of the RNN Classifier

Hyperparameters of the RNN Classifier	Data Type	Default Values	Description
Size of RNN	INT	256	This parameter records the size of the RNN hidden state. The rule of thumb for the RNN hidden state size is usually between the size of the input and size of the output layers.
Number of Layers for RNN	INT	1	This parameter records the number of layers in the RNN. One hidden layer is sufficient for the majority of the problems. For improving the performance additional layers can be added but it is not guaranteed.
Model of RNN	STRING	lstm	There are different RNN models such as Vanilla RNN (rnn), Gated Recurrent Unit (gru), Long Short-Term Memory (lstm), etc.
Batch size	INT	50	This parameter records the batch size. Let's say there are 1050 training samples, and the batch size is set equal to 100. In this case the algorithm takes the first 100 samples (from 1st to 100th record) from the training dataset and starts training the network. Next it takes the second 100 samples (from 101st to 200th record) and starts training the network again. This procedure is repeated until all the records in the training sample have been used for training the network.
Sequence Length	INT	25	This parameter records the sequence length for the input data used for training the RNN. To find the optimal value of the sequence length cross-validation can be performed either using the grid search or the Bayesian optimization. Sequence length doesn't really affect the model training performance.
Number of Epochs	INT	25	One epoch is equal to one forward pass and one backward pass of all the training samples. To find the optimal number of epochs, first calculate the RMSE of the training and test data for each epoch with different number of maximum epochs. This prevents the model from overfitting and gives an approximated range of epochs to start with. This method can be repeated by maintaining a constant epoch (previously selected) and testing the model with different number of neurons.
Gradient Clipping Value	FLOAT	5.0	When gradients are being propagated back in time, they can vanish as they are continuously multiplied by a number less than one. This is called the vanishing gradient problem. On the other hand, gradients can also explode. Gradient explodes when they get exponentially large from being multiplied by numbers larger than 1. Gradient clipping clips the gradients between two numbers to prevent them from getting too small or too large.

(Continued)

TABLE 10.1 (CONTINUED)

Hyperparameters of the RNN Classifier	Data Type	Default Values	Description
Learning Rate	FLOAT	0.002	This parameter records the RNN learning rate. If the learning rate is low, then training is more reliable, but to achieve optimization it will take a lot of time because it takes more time to minimize the loss function. If the learning rate is too high, then the training may not converge or even diverge. In this case the weight changes can be so huge that the optimizer might overshoot the minimum and make the loss worse.
Decay Rate	FLOAT	0.97	The decay rate is represented as *rmsprop*. After each iteration, the weights are multiplied by a factor slightly less than 1. This prevents the network weights from growing too large and can be seen as a gradient descent on a quadratic regularization term. Weight decay specifies the regularization in the neural network.
GPU Memory	FLOAT	0.666	This parameter records the percentage of GPU memory to be allocated to this process.
Dropout Rate	FLOAT	0.02	The default interpretation of the dropout hyperparameter is the probability of training a given node in a layer, where 1.0 means no dropout, and 0.0 means no outputs from the layer. A good value for dropout in a hidden layer is between 0.5 and 0.8. Input layers use a larger dropout rate, such as of 0.8.

To begin with the following libraries need to be installed in R [6]

- *Keras*
- *Tensorflow*

See the R code below for installing the *Keras* and *Tensorflow* packages

```
install.packages("keras")
install.packages("tensorflow")
```

Now that the *tensorflow* package has been downloaded you will need to load the library and install the *tensorflow* in your computer (see the R code below)

```
library(tensorflow)
install_tensorflow()
```

Now load the *keras* library

```
library(keras)
```

The next step is to load the Reuters dataset. The tokens (words) in the Reuters dataset are ranked by how often they occur (in the training set) and the *max_features* parameter controls how many distinct tokens (words) will be used in this modeling phase. In this example, we will use all the tokens. The *maxlen* parameter is used to control the length of the input sequences (input data) provided to the model. Here, we will choose to have the same length for all the input sequences. If the sequences are longer than the *maxlen* variable, then they are truncated; if they are shorter, then padding will be performed to make the length equal to the value of the *maxlen* variable. Since the *maxlen* variable is set to 150 here the LSTM model expects 150 tokens as input per instance (see the R code below)

```
set.seed(42)
word_index <- dataset_reuters_word_index()
max_features <- length(word_index)
maxlen <- 150
skip_top = 0

reuters <- dataset_reuters(num_words = max_features,skip_top =
  skip_top)
c(c(x_train, y_train), c(x_test, y_test)) %<-% reuters
x_train <- pad_sequences(x_train, maxlen = maxlen)
x_test <- pad_sequences(x_test, maxlen = maxlen)
x_train <- rbind(x_train,x_test)
y_train <- c(y_train,y_test)
table(y_train)
```

Finally, after merging the training and test sets, the distribution of the *y* variable is visualized. The distribution of the number of instances (not shown here) across 46 classes suggests that the dataset is highly imbalanced. In the next step, we will convert our text classification problem to a binary classification problem. The instances with the class label 3 will be re-labeled as 1 and the remaining instances with class labels of 0,1,2,4,…,45 will be labeled as 0.

```
y_train[y_train!=3] <- 0
y_train[y_train==3] <- 1
```

To generate the training data, execute the following R code

```
table(y_train)
```

The resultant distribution of the dataset is shown in Figure 10.7.

Here, it can be seen that there are 7,256 instances of class 0 and 3,972 instances of class 1.

The next section of the code builds the model. Note the use of the hyperparameters here including the *dropout rate*, *model of RNN*, *activation function*, choice of the *optimizer*, and the choice of the *loss function*. The batch size has been set to 32 and the number of epochs is set to 10. The validation split should be a number between 0 and 1 and we have chosen it to be 0.9 here. The other important parameters for the model are max length = 150, the size of the embedding layer = 32, and as mentioned before the model is trained for 10 epochs (see the R code below). The summary of the model is shown in Figure 10.8.

```
model <- keras_model_sequential() %>%
  layer_embedding(input_dim = max_features, output_dim =
  32,input_length = maxlen) %>%
  layer_dropout(rate = 0.25) %>%
  layer_lstm(128,dropout=0.2) %>%
  layer_dense(units = 1, activation = "sigmoid")

model %>% compile(
  optimizer = "rmsprop",
  loss = "binary_crossentropy",
  metrics = c("acc")
)
summary(model)
history <- model %>% fit(
  x_train, y_train,
  epochs = 10,
  batch_size = 32,
  validation_split = 0.9
)
```

The output for the model training is shown in Figure 10.9. Here, it can be seen that the best validation accuracy was after epoch 5, which is 92.66%.

```
y_train
   0    1
7256 3972
```

FIGURE 10.7 Distribution of training data for Reuters dataset.

```
Model: "sequential_2"
_________________________________________________________________
Layer (type)                 Output Shape              Param #
=================================================================
embedding_2 (Embedding)      (None, 150, 32)           991328
dropout_2 (Dropout)          (None, 150, 32)           0
lstm_2 (LSTM)                (None, 128)               82432
dense_2 (Dense)              (None, 1)                 129
=================================================================
Total params: 1,073,889
Trainable params: 1,073,889
Non-trainable params: 0
_________________________________________________________________
```

FIGURE 10.8 LSTM model specifications.

```
Epoch 1/10
36/36 [==============================] - 12s 284ms/step - loss: 0.5193 - acc: 0.7888 - val_loss: 0.4335 - val_acc: 0.8444
Epoch 2/10
36/36 [==============================] - 10s 288ms/step - loss: 0.3820 - acc: 0.8788 - val_loss: 0.3665 - val_acc: 0.8738
Epoch 3/10
36/36 [==============================] - 12s 332ms/step - loss: 0.2748 - acc: 0.9082 - val_loss: 0.3008 - val_acc: 0.9076
Epoch 4/10
36/36 [==============================] - 11s 325ms/step - loss: 0.2344 - acc: 0.9207 - val_loss: 0.2559 - val_acc: 0.9203
Epoch 5/10
36/36 [==============================] - 11s 325ms/step - loss: 0.2623 - acc: 0.9242 - val_loss: 0.2544 - val_acc: 0.9266
Epoch 6/10
36/36 [==============================] - 12s 327ms/step - loss: 0.1619 - acc: 0.9581 - val_loss: 0.2801 - val_acc: 0.9261
Epoch 7/10
36/36 [==============================] - 12s 339ms/step - loss: 0.1416 - acc: 0.9661 - val_loss: 0.3712 - val_acc: 0.9062
Epoch 8/10
36/36 [==============================] - 12s 326ms/step - loss: 0.1311 - acc: 0.9688 - val_loss: 0.2527 - val_acc: 0.9139
Epoch 9/10
36/36 [==============================] - 11s 324ms/step - loss: 0.1072 - acc: 0.9750 - val_loss: 0.2445 - val_acc: 0.9197
Epoch 10/10
36/36 [==============================] - 12s 333ms/step - loss: 0.1339 - acc: 0.9697 - val_loss: 0.2318 - val_acc: 0.9263
```

FIGURE 10.9 Model training output for different epochs.

SUMMARY

The focus of this chapter was to introduce RNN architecture and the different hyperparameters that can be tuned to effectively perform text classification. To this end only the LSTM architecture was discussed as a RNN structure. There are other RNN structures such as the Gated Recurrent Units (GRU), Bidirectional LSTM, Stacked Bidirectional LSTM, etc. The other RNN structures are not discussed here since they are either similar to LSTM or use the LSTM as a building block.

The RNNs are best suited for the text classification as text is naturally sequential and the RNN, particularly LSTM, takes a natural approach to remember the long-term dependencies in the text sequences.

SIDEBAR 1 VANISHING GRADIENT PROBLEM

The vanishing gradient problem occurs when many layers using certain activation functions are added to the neural networks. In such a case the gradients of the loss function approach to zero which makes the network hard to train [7].

For example, consider the sigmoid function $s(x)=\frac{1}{1+e^{-x}}$. This function compresses a large input space into a small input space, i.e., between 0 and 1. Therefore, a large change in the input of the sigmoid function will cause a small change in the output. Hence, the derivative becomes small.

When the network is shallow with only a few layers, then it is not a significant issue but when more layers are used, it can cause the gradient to be too small for the training to work effectively. Generally, the gradients of neural networks are found using backpropagation. During the backpropagation the derivatives of the network are determined by moving through a single layer at a time starting from the final layer to the initial layer.

According to the chain rule, the derivatives of each layer are multiplied from the final layer to the initial layers. In this case when the hidden layers use an activation like the sigmoid function the small derivatives get multiplied together. Thus, the gradient decreases exponentially as we move from the final layer to the initial layer. Here, a small gradient means that the weights and biases of the initial layers will not be updated effectively when the training session is iterated. Since the initial layers are most crucial for recognizing the core elements of the input data, the overall accuracy of the entire network is severely challenged [7].

SIDEBAR 2 THE GRADIENT DESCENT ALGORITHM

Gradient Descent is an optimization algorithm to find the minimum of a function (typically a cost function under constraints) using the first order derivative where the objective is to minimize the cost function $J(w,b)$ with regards to weights and bias, i.e., w and b. Here, the cost function $J(w,b)$ is the metric also known as the loss function that determines how well or poorly a machine learning algorithm performed with regards to the actual training output and the predicted output. Mathematically, $J(w,b)$ is expressed as [2, 7]

$$J(w,b) = \frac{-1}{m}\sum_{i=1}^{m} y_{(i)}\log\hat{y}_{(i)} + \left(1-y_{(i)}\right)\log\left(1-\hat{y}_{(i)}\right) \quad (10.1)$$

where m is the size of the training dataset, $y_{(i)} \in [0,1]$ is the vector of outputs and $\hat{y}_{(i)}$ is the vector of the predicted outputs. Here, it is important to note that the classification task is non-convex and as a result we need to use the cross-entropy loss such as the $J(w,b)$ as a cost function.

In an effort to minimize the $J(w,b)$ after numerous iterations, the following steps are included

$$w - \alpha \times \frac{\partial J(w,b)}{\partial w} \rightarrow w$$

$$b - \alpha \times \frac{\partial J(w,b)}{\partial b} \rightarrow b$$

Here, α used in the above equations is referred to as the *learning rate*. The learning rate is the rate at which the learning agent learns the new knowledge. Thus α is the hyperparameter that needs to be assigned as a scalar value or as a function of time. In each iteration the values of w and b are updated until the value of the cost function reaches an acceptable minimum value [2, 7].

The Gradient Descent algorithm is all about moving down the slope of the curve which is represented using the cost function $J(w,b)$ with regards to the parameters. The gradient or the slope gives the direction of the increasing or decreasing slope if it is positive or negative, respectively. The negative sign in Equation 10.1 is multiplied with the slope to indicate that we need to move in the opposite direction. The descent is controlled by choosing an optimal value for the learning rate α. If α is very small, then the convergence will take more time and while it is very high it will overshoot and miss the minimum and diverge leading to a large number of iterations [2, 7].

A well-known disadvantage of Gradient Descent is that it can get "trapped" in local minima, forgoing the optimal solution. Combining Gradient Descent with some form of randomization, however, can help avoid being trapped in local minima.

EXERCISE

1. One of the classifier listed below has the ability to perform sequence-based learning
   A. ANN
   B. RNN
   C. DNN
   D. CNN
   E. CNN and RNN

2. Download the Reuters dataset (discussed in lesson 10). Choose instances from this dataset pertaining to any 5 different categories (out of the 46 categories). First, summarize this new dataset. Second, use the RNN model based on the LSTM architecture to determine the 5 categories of text using all the tokens. Set the *maxlen* variable to 150. For validation testing, split the dataset into a ratio of 80:20. Clearly highlight the choices for the hyperparameters used in the modeling of the LSTM and summarize the performance of the classifier.

3. Repeat all the instructions provided in question 2. This time set the *maxlen* variable to 250. Does increasing the value of the *maxlen* variable improve the performance of the classifier?

4. What is the vanishing gradient problem? How does LSTM mitigate the vanishing gradient problem?

5. The LSTM architecture has a chain structure that contains four neural networks and different memory blocks called ___________
   A. Input gate
   B. Forget gate
   C. Cells
   D. Output gate
   E. None of the above

6. Discuss the role and importance of the different gates within a LSTM cell?

REFERENCES

1. Biswal, A. (2022). "Recurrent Neural Network (RNN) Tutorial: Types, Examples, LSTM and More", retrieved from Recurrent Neural Network (RNN) Tutorial: Types and Examples [Updated] | Simplilearn, retrieved on June 18, 2022.
2. Dutta, S. (2018). *Reinforcement Learning with TensorFlow*. Packt Publishing Ltd., ISBN 978-1-78883-572-5.
3. Narwekar, A., Pampari, A. (2016). "Recurrent Neural Network Architectures", retrieved from Recurrent Neural Network Architectures (illinois.edu), retrieved on June 18, 2022.
4. Jayawardhana, S. (2020). "Sequence Models & Recurrent Neural Networks (RNNs)", retrieved from https://towardsdatascience.com/sequence-models-and-recurrent-neural-networks-rnns-62cadeb4f1e1, retrieved on June 18, 2022.
5. Eliasy, A., Przychodzen, J. (2020). The Role of AI in Capital Structure to Enhance Corporate Funding Strategies. *Array*, 6, 100017, ISSN 2590-0056, https://doi.org/10.1016/j.array.2020.100017.1–13.
6. Browniee, J. (2016). "Sequence Classification with LSTM Recurrent Neural Networks in Python with Keras", retrieved from Sequence Classification with LSTM Recurrent Neural Networks in Python with Keras (machinelearningmastery.com), retrieved on June 18, 2022.
7. Wang, Chi-F. (2019). "The Vanishing Gradient Problem – The Problem, Its Causes, Its Significance, and Its solution", retrieved from https://towardsdatascience.com/the-vanishing-gradient-problem-69bf08b15484, retrieved on June 18, 2022.

11 Recommender Systems for Predictive Analytics

In earlier chapters we discussed the basics of the neural and deep neural networks. Here, we will discuss recommendation systems that are widely used in our day-to-day online activities and many personal digital assistants. The rise in popularity of YouTube, Amazon, Netflix, Facebook, and many other such web services has given impetus for developing efficient recommender systems. A *recommender system* (RS) or *recommender engine* (RE) is an algorithm designed to understand user behaviors and suggest relevant items or actions to the users [1]. Recommender systems are famously known for explaining user behavior and their buying patterns.

As an example, consider shopping on the Amazon website. How does Amazon recommend books based on ones you have previously bought? This RE is an example of a large-scale machine learning algorithm. It mines all the previously collected and stored transaction data about book purchases and the characteristics of the buyers, and looks for behavioral patterns. Once the patterns are identified, it is assigned a weight, which is used to make recommendations. In Chapter 5 we discussed the Market Basket technique and highlighted how the patterns are mined on transactional data.

RE's are used everywhere. In Facebook/LinkedIn, the RE recommends the most probable people one may like to befriend or add them as professionals in their network. Ancestry.com and other genealogical research sites use REs to suggest possibly family matches. Therefore, both the RS and REs are a driving force for these companies. RE can also be used in computer games (e.g., to suggest possible actions). Sidebar 2 describes an application of a RE to the game "20 Questions."

Generally, the concept of both RE and RS is to predict what people might like and to uncover relationships between items/products to aid in the discovery process. In that respect, they are both similar to a search engine. However, a search engine shows results only when the user requests something while a RE/RS is more proactive in nature. It presents users with relevant content that is not necessarily requested by the user.

RS and RE are really critical in most of the industries as they can generate a huge amount of income and profits. In this chapter, we will discuss the different paradigms of RS/RE. For each of them, we will briefly show how they work, their theoretical basis, and their strengths and weaknesses. In addition, a case study–based example will be provided to compare the different RS/RE approaches using the R scripting language.

DOI: 10.1201/9781003278177-11

CONTENT-BASED FILTERING

Content-based filtering uses the content or attributes of the item, together with a notion of similarity between the two pieces of content, to generate similar items with respect to the given item. For these types of approaches, the idea is to link the user preferences with the item attributes [1].

For example, for a movie, attributes include genre, cast, storyline, and so on. Recommendations to a user for a movie to watch would match these attributes, and then select from the set of movies those with the highest user ratings from other viewers (e.g., movies rated at least 4.5 out of 5 stars). The advantage of this approach is that when a new item is added, it can be recommended to a user if it matches the attribute preferences in their profile.

Content-based methods can suffer when limited content is available about the user preferences or about the item in question. These approaches can result in non-unique recommendations or poor recommendations. Cosine similarity is a well-known technique that can be used for recommending items to the user. Let's look at this technique and illustrate this concept using an example.

COSINE SIMILARITY

Cosine similarity is a metric that is used to measure how similar things are irrespective of their size. In a mathematical sense it measures the cosine of the angle between two vectors that is projected in a multidimensional space [1]. For an arbitrary thing, each dimension represents a characteristic of that thing.

For example, suppose we are looking to determine the similarity between two dogs along three dimensions: height, weight, and age. For simplicity let's assume the range of each dimension is a positive real number. Now consider dog 1, Arrow, whose shoulder to ground height (in meters), weight (in kg) and age (in years) is represented by the vector A = (0.2, 9.1,3.2). Now consider a second dog, Boomerang, who is characterized by the vector B = (0.68, 27.2, 6.0).

The merit behind using this metric is that even if two similar things are separated far apart in the Euclidean plane the chances are that they may still be oriented closer together or be similar to each other. This metric measures the angle between the two vectors and is interpreted as smaller the angle, higher the cosine similarity. Consider the vectors $\underset{A}{\rightarrow}$ and $\underset{B}{\rightarrow}$ in a two-dimensional plane as shown in Figure 11.1. The Cosine similarity is defined as

$$\cos\theta = \frac{\sum_{i=1}^{n} A_i B_i}{\sqrt{\sum_{i=1}^{n} A_i^2}\sqrt{\sum_{i=1}^{n} B_i^2}}$$

where A_i and B_i are components of the vectors A and B.

As a quick example, let's consider the cosine similarity for our dogs Arrow and Boomerang:

A = (0.2, 9.1,3.2), B = (0.68, 27.2, 6.0).

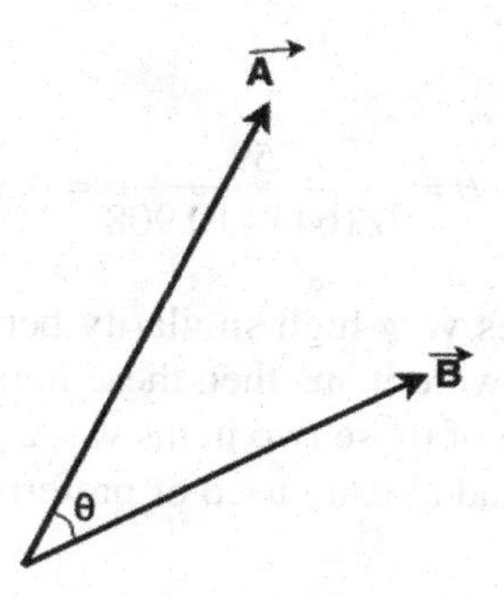

FIGURE 11.1 Vectors A and B in a multidimensional plane.

$$\sum_{i=1}^{3} A_i B_i = ((0.2*0.68)+(9.1*27.2)+(3.2*6.0)) = 266.86$$

$$\sqrt{\sum_{i=1}^{3} A_i^2} = \sqrt{0.2^2 + 9.1^2 + 3.2^2} = 93.09$$

$$\sqrt{\sum_{i=1}^{3} B_i^2} = \sqrt{0.68^2 + 27.2^2 + 6.0^2} = 776.30$$

Therefore, $\cos\theta = \dfrac{266.856}{93.09 * 776.302} = 0.307$

A value of 0.307 indicates a low similarity between the two vectors, as the highest possible value of cosine similarity is 1. This finding makes sense since clearly Arrow is much smaller and younger than Boomerang.

For most applications, usually there are many more than three dimensions characterizing the things. Let's look at a higher dimensional example of cosine similarity. Assume that in a 7-dimensional plane, there are two vectors $\underset{A}{\rightarrow} = (2,2,1,0,2,3,4)$ and $\underset{B}{\rightarrow} = (1,2,3,0,5,8,4)$ and we would like to compute the cosine similarity between them.

$$\sum_{i=1}^{7} A_i B_i = ((2*1)+(2*2)+(1*3)+(0*0)+(2*5)+(3*8)+(4*4)) = 59$$

$$\sqrt{\sum_{i=1}^{7} A_i^2} = \sqrt{2^2 + 2^2 + 1^2 + 0^2 + 2^2 + 3^2 + 4^2} = 6.1644$$

$$\sqrt{\sum_{i=1}^{7} B_i^2} = \sqrt{1^2 + 2^2 + 3^2 + 0^2 + 5^2 + 8^2 + 4^2} = 10.908$$

Therefore,

$$\cos\theta = \frac{59}{6.1644 * 10.908} = 0.877$$

A value of 0.877 indicates very high similarity between the two vectors. In this case, if the vectors *A* and *B* were items then these items would be recommended to the user given that either one of those two items was a previous preference indicated by the user, i.e., if the user had already used or preferred *A* then *B* would be recommended or vice-versa.

COLLABORATIVE FILTERING

The *collaborative filtering* approach is a form of wisdom-of-the-crowd. This approach generates an estimated preferences of users for items based on the set of preferences of many users for the item(s) in question. In short, this technique estimates the preferences of users for the items which have not yet been rated/reviewed. It works on the notion of similarity. This methodology determines similar users and their ratings by mining similar preferences exhibited by other users. For example, If the User A and User B have indicated similar rating for a movie X then they will give a similar rating for the movie Y [1, 2].

Collaborative filtering provides many advantages over content-based filtering. In collaborative filtering the content of the items does not necessarily tell the whole story. Even when no information on an item is available, we can still predict the item rating without waiting for a user to rate it. The collaborative technique also has the capability to capture the change of user interests over the period of time. In addition to that the collaborative filtering is also capable of capturing the inherent subtle characteristics, i.e., if most users buy two unrelated items, then it is highly likely that another user who shares similar interests with other users is also likely to buy that unrelated item.

There are four collaborative-based filtering approaches [1]:

1. User-based collaborative filtering (UBCF)
2. Item-based collaborative filtering (IBCF)
3. Singular value decomposition (SVD)
4. Principal components analysis (PCA)

We discussed SVD and PCA in Chapter 2. Here, we'll focus our discussion on UBCF and IBCF.

USER-BASED COLLABORATIVE FILTERING (UBCF)

In the UBCF, the objective is to find the missing ratings for a user by first identifying a neighborhood of similar users [1, 2]. Once the similar users have been identified, aggregation of their ratings is formed for prediction. The number of users in a neighborhood is determined by either using the KNN algorithm, i.e., by setting the value of k, or by using a similarity measure with a minimum threshold. The two similarity measures commonly used are the *pearson correlation coefficient* and *cosine similarity*. The weakness of the UBCF algorithm is that the entire dataset has to be in the memory, and it is very time consuming [1].

Let's see an example to illustrate the UBCF algorithm. Consider the user-item dataset containing the rating of four different films given by six different users (see Figure 11.2).

Notice that the last entry, i.e., rating for the film 4 by user 6 is missing in Figure 11.2. The objective here is to use the UBCF algorithm to determine the missing value of the rating. Using $k = 1$, we can see that both "User 3" and "User 6" have provided a rating of 5 for "Film 2." Therefore, it is likely that the "User 6" would rate "Film 4" as 4. In a similar manner, using $k = 1$, we can see that both "User 5" and "User 6" have provided a rating of 1 for "Film 1." Therefore, it is likely that "User 6" would rate "Film 4" as 1 or 4.

Higher values of k are not possible as there are no two users who have rated the movies with same points. To summarize, the UBCF algorithm looks for two different users who have provided similar ratings for k different items.

Item-Based Collaborative Filtering (IBCF)

The IBCF algorithm uses the similarity between the items and not the users to make a recommendation. The assumption is that the user will prefer items that are similar to the other items that they like. The strategy is to build a model by calculating a pair-wise similarity matrix of all the items. The most commonly used similarity measures are *pearson correlation* (see Chapter 3) and *cosine similarity*. The size of the similarity matrix can be reduced by specifying the value of k, i.e., to retain only the k-most similar items. However, limiting the size of the neighborhood significantly reduces the accuracy thus resulting in poor performance of the algorithm [1, 2].

In Figure 11.3, Let's assume that the last entry, i.e., rating for the film 4 by user 6 is missing. Using $k = 1$, we can see that "User 5" has provided a rating of 1 for "Film 2" and "Film 4."

Therefore, there is a high chance that "User 6" will rank "Film 4" as 5 because "User 6" has already ranked "Film 2" as 5.

	Film 1	Film 2	Film 3	Film 4
User 1	3	5	3	4
User 2	5	2	5	3
User 3	5	5	1	4
User 4	5	1	5	2
User 5	1	1	4	1
User 6	1	5	2	4

FIGURE 11.2 User-item dataset containing the ratings for films by users.

	Film 1	Film 2	Film 3	Film 4
User 1	3	5	3	4
User 2	5	2	5	3
User 3	5	5	1	4
User 4	5	1	5	2
User 5	1	1	4	1
User 6	1	5	2	

FIGURE 11.3 User-item dataset containing the ratings for film with missing data.

HYBRID RECOMMENDATION SYSTEMS

A *hybrid recommendation system* (HRS) combines the content and collaborative filtering method. The combination of the collaborative and content-based filtering together can help in overcoming the shortcomings faced by each technique separately and can prove to be more effective in some cases [3–5].

HRS approaches can be implemented in various ways. For example, one can combine the content and collaborative-based methods to generate predictions separately and then combine the prediction [3] (see Figure 11.4).

Alternatively, the capabilities of both the approaches can be added successively. Several studies have advocated for the potentiality of this approach and suggest that more accurate recommendations can be generated using the HRS.

There are 7 approaches to building the HRS [4, 5]. These are as follows.

Weighted: In the weighted recommendation system the outputs are taken from each of the models and the results are combined using static weightings, i.e., the weights do not change across the training and test sets. For example, one can combine the content-based model and an IBCF model and assign a 50% weightage toward each model for the final prediction (see Figure 11.5).

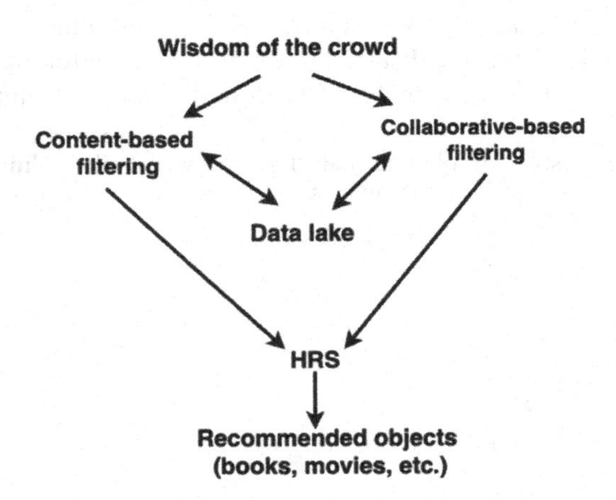

FIGURE 11.4 Depiction of an HRS approach. (Source: Figure adapted from Verma, Y. (2021). "A Guide to Building Hybrid Recommendation Systems for Beginners." Retrieved from https://analyticsindiamag.com/a-guide-to-building-hybrid-recommendation-systems-for-beginners/, retrieved on July 9, 2022.)

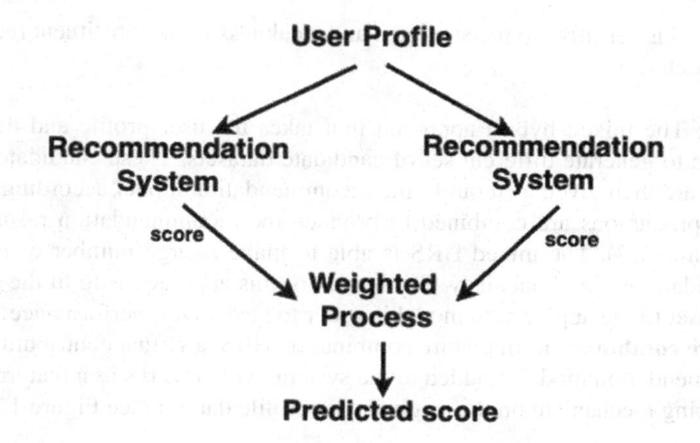

FIGURE 11.5 The Weighted HRS process. (Source: Figure adapted from Chiang, J. (2021). "7 types of Hybrid Recommendation System" retrieved from https://medium.com/analytics-vidhya/7-types-of-hybrid-recommendation-system-3e4f78266ad8, retrieved on July 9, 2022.)

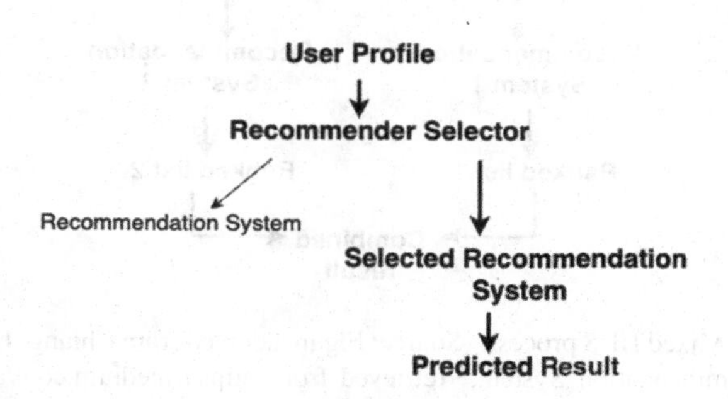

FIGURE 11.6 The switching HRS process. (Source: Figure adapted from Chiang, J. (2021). "7 types of Hybrid Recommendation System" retrieved from https://medium.com/analytics-vidhya/7-types-of-hybrid-recommendation-system-3e4f78266ad8, retrieved on July 9, 2022.)

The benefit of using the weighted HRS is that one can consider multiple models on the dataset to support the recommendation process in a linear way.

Switching: The switching HRS selects a single recommendation system based on the given situation. The model is commonly used when the dataset is sensitive at the item-level. A recommender selector criterion is set based on the user profile or other features. The switching HRS approach introduces an additional layer on top of the recommendation model, which select the appropriate model to use (see Figure 11.6).

This HRS is sensitive to the strengths and weakness of its constituent recommendation model(s).

Mixed: The mixed hybrid approach first takes the user profile and the features to generate different set of candidate datasets. These candidate datasets are then given as input to the recommendation model accordingly and the predictions are combined to produce the recommendation result (see Figure 11.7). The mixed HRS is able to make a large number of recommendations simultaneously. The benefit of this approach is to fit the partial dataset to the appropriate model in order to have better performance.

Feature combination: In feature combination HRS, a virtual contributing recommendation model is added to the system, which works as a feature engineering mechanism on the original user profile dataset (see Figure 11.8).

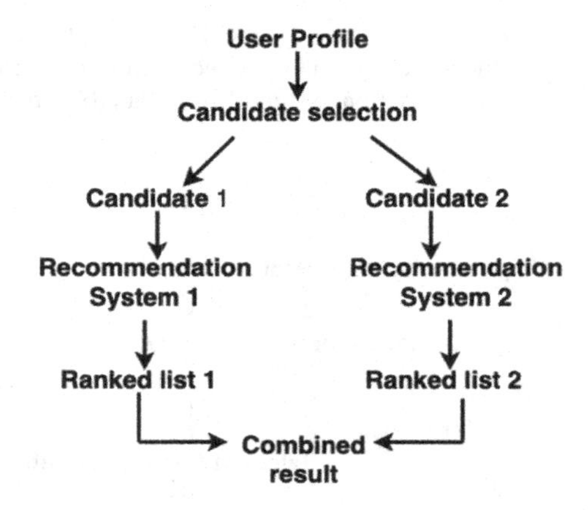

FIGURE 11.7 Mixed HRS process. (Source: Figure adapted from Chiang, J. (2021). "7 types of Hybrid Recommendation System," retrieved from https://medium.com/analytics-vidhya/7-types-of-hybrid-recommendation-system-3e4f78266ad8, retrieved on July 9, 2022.)

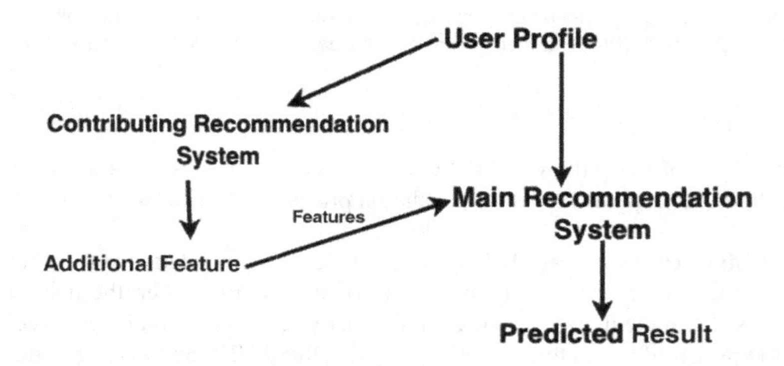

FIGURE 11.8 Feature combination HRS process. (Source: Figure adapted from Chiang, J. (2021). "7 types of Hybrid Recommendation System" retrieved from https://medium.com/analytics-vidhya/7-types-of-hybrid-recommendation-system-3e4f78266ad8, retrieved on July 9, 2022.)

For example, the features of a collaborative recommendation model can be injected into a content-based recommendation model. This HRS model is capable enough to consider the collaborative data from the sub system while relying on one model exclusively.

Feature augmentation: In this HRS a contributing recommendation model is employed to generate a rating or classification of the user/item profile. This is further used in the main recommendation system to produce the final predicted result. The feature augmentation HRS improves the performance of the core system without changing the main recommendation model. For example, the user profile dataset can be enhanced with the determination of the association rules (see Figure 11.9).

Here, the content-based recommendation model experiences improvement with the addition of the augmented dataset.

Cascade: The Cascade HRS defines a strict hierarchical structure (see Figure 11.10).

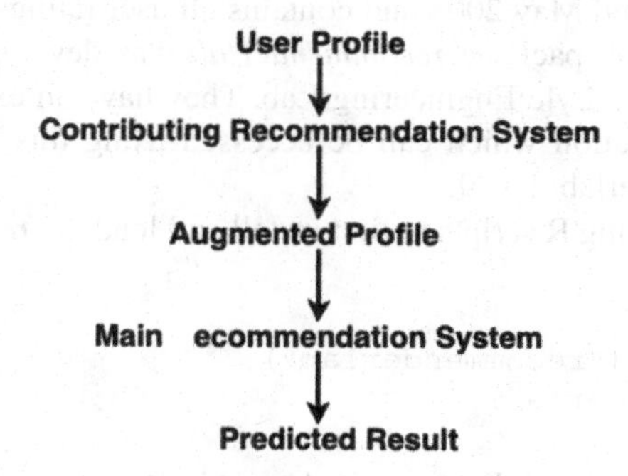

FIGURE 11.9 Feature augmentation HRS process. (Source: Figure adapted from Chiang, J. (2021). "7 types of Hybrid Recommendation System" retrieved from https://medium.com/analytics-vidhya/7-types-of-hybrid-recommendation-system-3e4f78266ad8, retrieved on July 9, 2022.)

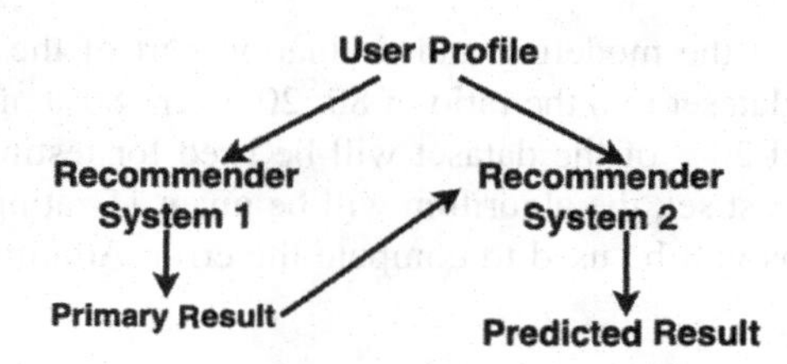

FIGURE 11.10 Cascade HRS process. (Source: Figure adapted from Chiang, J. (2021). "7 types of Hybrid Recommendation System," retrieved from https://medium.com/analytics-vidhya/7-types-of-hybrid-recommendation-system-3e4f78266ad8, retrieved on July 9, 2022.)

The main recommendation system produces the primary result, which is then used by the secondary model to resolve minor issues in the primary result. For example, breaking ties in the scoring.

Meta-level: The Meta-level HRS is similar to the feature augmentation HRS. The contributing model in the meta-level HRS provides an augmented dataset to the main recommendation model. However, unlike in the feature augmentation HRS, the meta-level HRS replaces the original dataset with a learned model from the contributing model, which then becomes the input to the main recommendation model.

Examples of Using Hybrid Recommendation Systems

Now that we have discussed the different RE and RS let's explore their usage in the *Jester5k* dataset obtained from the *recommenderlab* package. Here, we'll use the R language for scripting. The objective of this exercise will be to compare the performance of the different RE and RS.

The *Jester5k* dataset consists of 5,000 ratings on 100 jokes which were sampled from the Jester Online Joke Recommender System. This dataset was collected between April 1999 and May 2003 and contains all user ratings on at least 36 jokes. On the other hand, the package *recommenderlab* was developed by the Southern Methodist University's Lyle Engineering Lab. They have an excellent website with supporting documentation which can be accessed using this link https://lyle.smu.edu/IDA/recommenderlab/ [5, 6].

Execute the following R scripts to first install and load the *recommenderlab* package [7]

```
install.packages("recommenderlab")
library(recommenderlab)
```

The next step will be to load the *Jester5k* dataset in R

```
data(Jester5k)
Jester5k
```

Upon executing the above scripts, one should be able to see the output as shown in Figure 11.11.

Now let's focus on the modeling and evaluation part of the different RE algorithms. We split the dataset into the ratio of 80::20 where 80% of the dataset will be used for training and 20% of the dataset will be used for testing. Here, we'll also specify that, for the test set, the algorithm will be given 15 ratings. This means that the other rating items will be used to compute the error. Additionally, we will also

```
5000 x 100 rating matrix of class 'realRatingMatrix' with 363209 ratings.
```

FIGURE 11.11 *Jester5k* dataset summary.

specify the threshold for a good rating which is set to be greater than or equal to 5. To do so execute the following R scripts

```
set.seed(123)
e<- evaluationScheme(Jester5k, method="split",train=0.8,
  given=15, goodRating=5)
```

Now that we have the training and test dataset, let's begin to model and evaluate the different REs namely the UBCF, IBCF, and SVD. In order to build and test the RE's, we will use the same function, *Recommender()* by merely changing the specification for each technique. See the following R scripts below

```
ubcf <- Recommender(getData(e,"train"), "UBCF")
ibcf <- Recommender(getData(e,"train"), "IBCF")
svd <- Recommender(getData(e, "train"), "SVD")
```

Now, using the *predict()* function, we can obtain the predicted ratings for the 15 items in the test data across each of the algorithms. Execute the R script below

```
user_pred <- predict(ubcf, getData(e, "known"), type =
  "ratings")
item_pred <- predict(ibcf, getData(e, "known"), type =
  "ratings")
svd_pred <- predict(svd, getData(e, "known"), type =
  "ratings")
```

In order to examine the error in the predictions on the test data the package provides the *calcPredictionAccuracy()* function. This function will output three metrics namely RMSE, MSE, and MAE across all the REs (see Sidebar 1 for more details about these metrics). Now execute the following R scripts.

```
P1 <- calcPredictionAccuracy(user_pred, getData(e,"unknown"))
P2 <- calcPredictionAccuracy(item_pred, getData(e, "unknown"))
P3 <- calcPredictionAccuracy(svd_pred, getData(e, "unknown"))
```

To examine the errors of the predictions for all the RE's execute the R scripts provided below

```
error <- rbind(P1, P2, P3)
rownames(error) <- c("UBCF", "IBCF", "SVD")
error
```

From the output we can see that the SVD algorithm slightly outperforms the UBCF and IBCF algorithm (see Figure 11.12). The RE algorithm that results in a lower value for all the three metrics namely RMSE, MSE and MAE, is the best. This conclusion, as such, cannot be generalized because on a different partition of this dataset the performance of the REs can differ. However, this exercise serves as a demonstration for how the different RE and RS can be compared against each other.

```
> error
         RMSE      MSE      MAE
UBCF 4.643612 21.56313 3.646172
IBCF 4.527517 20.49841 3.473779
SVD  4.317032 18.63676 3.398328
```

FIGURE 11.12 Performance metrics for all the REs.

SUMMARY

This chapter focuses on introducing the recommendation systems for predictive analytics. Here, we have discussed the algorithms for the content-based, collaborative approaches for designing different hybrid recommendation systems. Despite being robust the content-based filtering approaches suffer from making inaccurate recommendations for new or inactive users. Even though collaborative approaches address the limitations of content-based filtering it suffers from data sparsity for less popular items with only few ratings. The hybrid-based filtering technique on the other hand is essentially an ensemble approach that implements both the content-based and collaborative methods separately and combines their predictions. To illustrate the merits of the hybrid approach, this chapter presents a case study on the Jester 5k dataset.

SIDEBAR 1 MAE AND RMSE

To evaluate a prediction, one should compute the deviation of the prediction from the true value. This can be done using the metric *Mean Average Error* (MAE) which is given as

$$\text{MAE} = \frac{1}{|\mathcal{K}|} \sum_{(j,l)\in\mathcal{K}} |r_{jl} - \hat{r}_{jl}|$$

where $\mathcal{K}$ is the set of all user-item pairings (j,l) for which we have predicted ratings $\hat{r}_{jl}$ and also known rating r_{jl} which was not used for learning purpose by a recommender model.

The *Root Mean Square Error* (RMSE) penalizes larger errors much stronger than the MAE and thus is suitable for situations where small prediction errors are not very important. The RMSE is given by the expression

$$\text{RMSE} = \sqrt{\frac{\sum_{(j,l)\in\mathcal{K}} (r_{jl} - \hat{r}_{jl})^2}{|\mathcal{K}|}}$$

Consider the confusion matrix provided below

Actual/Predicted	Negative	Positive
Negative	a	b
Positive	c	d

The error measure *Mean Absolute Error* (MAE) is given by the expression

$$\text{MAE} = \frac{1}{N}\sum_{i=1}^{N} |\epsilon_i| = \frac{b+c}{a+b+c+d}$$

where $N = a+b+c+d$ is the total number of items which can be recommended and $|\epsilon_i|$ is the absolute error of each item.

SIDEBAR 2 THE 20-QUESTION GAME

"20 Questions" (20Q) is a game that involves two players, the "subject" (usually a human) and the "interrogator" (another human or a computer program). The role of the interrogator is to ask the subject a series of questions to determine a secret thing that the subject has chosen. The thing can be a physical object, a plant, animal, or even a concept, though there are usually limitations established. In its simplest form, the interrogator can ask up to 20 "yes" or "no" questions before they must guess the secret thing, though they may choose to guess before 20 questions have been asked.

Here's an example of the gameplay. Suppose the subject has chosen as the secret object a watermelon. The game proceeds with the interrogator's questions followed by the subject's answers:

It is something you can eat?
Yes

Does it have anything to do with salad?
No.

Does it roll?
Yes.

Is it shiny?
No.

Is it crunchy?
No.

Can it be used in a pie?
No.

Does it have a good smell?
Yes.

Is it striped?
Yes.

Does it weigh more than a duck?
Yes.

Does it burn?
No.

Is it smaller than a loaf of bread?
No.

Does it have leaves?
No.

Is it flexible?
No.

The interrogator now correctly guesses that the secret thing is a watermelon.

It's easy to see how this kind 20Q is a kind of recommender system. For example, imagine the subject wishes to purchase a book, but does not know which one. Using a series of questions (e.g., do you like fiction?) the recommender system can guide the purchase.

While 20Q was first implemented through decision tree logic, this approach is cumbersome and requires an extensive database of things. Modern implementations of 20Q (as do recommender systems) use neural networks. Over time, as the weights are adjusted, the system learns and improves its recommendations. These systems also support non-binary answers such as "unknown," "maybe," or even allow the subject to answer incorrectly.

20Q has an interesting history. It was a popular a parlor game in the mid-19th century and may have existed earlier. A radio show based on the game was broadcast in the 1940s and later appeared on television in the 1950s. In the 1970s a computer program that could only guess animals was developed and an electronic toy based on the game emerged shortly thereafter. A more substantial version of a 20Q program was popular in the 1980s, but all of the computer/electronic implementations of the game were essentially based on decision trees. Once NN were discovered, the game was quickly implemented there as a demonstration of AI's potential.

You can play a NN-based version of the game at: http://www.20q.net/. Some virtual assistants, such as Amazon's Alexa, also provide this game. A simple, electronic toy version of the game is also available, and it makes a great gift for inquisitive children, teens, and adults.

EXERCISE

1. The following are the predicted output of a classifier on a labeled dataset containing 12 instances. Construct (show) the confusion matrix and also compute

the following metrics: Accuracy, Sensitivity, False Negative Rate, Precision, and F1 score

Individual Number	1	2	3	4	5	6	7	8	9	10	11	12
Actual Classification	1	1	1	1	1	1	1	1	0	0	0	0
Predicted Classification	0	0	1	1	1	1	1	1	1	0	0	0

Source: Figure adapted from https://en.wikipedia.org/wiki/Confusion_matrix.

Consider the ratings of the user 1 to user 5 for 5 movies on a rating scale of 1–5 as shown below

	Movie 1	Movie 2	Movie 3	Movie 4	Movie 5
User 1	3	1	3	4	1
User 2	5	2	5	3	**y**
User 3	5	5	**x**	4	5
User 4	5	5	1	2	2
User 5	1	1	5	1	3

2. Using the UBCF and the k-NN algorithm where $k = 2$ the value of x should be
   A. 1
   B. 2
   C. 3
   D. 4
   E. 5

3. Using the IBCF and the k-NN algorithm where $k = 2$ the value of y should be
   A. 1
   B. 2
   C. 3
   D. 4
   E. 5

4. Determine the cosine similarity between two vectors $X = \{3,2,0,5\}$ and $Y = \{0,5,2,3\}$.

5. What type of recommendation systems/recommendation engines will you propose for researches to identify articles in the area of predictive analytics? Discuss the pros and cons of each recommendation system/engine.

6. Download the Jester 5k dataset (discussed in lesson 11) and extract a random sample containing only 75% of the instances from the original dataset. On this new dataset compare the performance of different recommendation systems including SVD, UBCF, and IBCF. Perform validation tests after splitting the new dataset in the ratio of 70:30 and report all the performance measures. Which recommendation system has the best performance in the new dataset?

REFERENCES

1. Dangeti, P. (2017). *Statistics for Machine Learning*, Packt Publishing Ltd., ISBN 978-1-78829-575-8.
2. Lesmeister, C. (2017). *Mastering Machine Learning with R*, Packt Publishing Ltd., ISBN 978-1-78728-747-1.
3. Burke, R. (2002). Hybrid Recommender Systems: Survey and Experiments. *User Model User-Adap Inter*, 12, 331–370. doi:10.1023/A:1021240730564.
4. Chiang, J. (2021). "7 types of Hybrid Recommendation System", retrieved from https://medium.com/analytics-vidhya/7-types-of-hybrid-recommendation-system-3e4f78266ad8, retrieved on July 9, 2022.
5. Verma, Y. (2021). "A Guide to Building Hybrid Recommendation Systems for Beginners", retrieved from https://analyticsindiamag.com/a-guide-to-building-hybrid-recommendation-systems-for-beginners/, retrieved on July 9, 2022.
6. Hahsler, M. (2022). "Recommenderlab: An R Framework for Developing and Testing Recommendation Algorithms", retrieved from https://cran.r-project.org/web/packages/recommenderlab/vignettes/recommenderlab.pdf, retrieved on July 9, 2022.
7. Chapman, C., Feit, E. M. (2015). *R for Marketing Research and Analytics*, Springer, ISBN 978-3-319-14436-8.

12 Architecting Big Data Analytical Pipeline

In this chapter, we discuss the Big Data technology landscape and analytics platform, introduce the lambda architecture, and discuss design strategies for building a customized Big Data pipeline using design patterns, and the associated pattern language.

Modern technologies have the capability to store, process, and analyze data in a scalable manner and thus are the core of any Big Data stack. The era of tables and records, which took over from file-based sequential storage, has now revolutionized to the era well-known as Big Data. In this era, we are able to harness the storage and computation power very well to support the day-to-day operations of multinational enterprises.

The era of relational database management systems (RDBMS) had survived for a long time until the era of 5 Vs (discussed in Chapter 1) emerged making the technology of previous era obsolete. The scaling of traditional RDBMS, at the computation power expected to process a huge amount of data with low latency, comes at a very high price. This reality led to the emergence of new technologies that have low cost, low latency, and are highly scalable. Modern technology deals with clusters containing hundreds and thousands of nodes, hurling, and churning several petabytes of data [1]. A few examples of key technologies include

Hadoop: Hadoop supports data storage and computations using a distributed framework of commodity hardware in a highly reliable and scalable manner. Hadoop distributes the data in chunks over different nodes in the cluster and then processes them concurrently across the nodes. The key components of Hadoop are mappers and reducers.

NoSQL: NoSQL is a tool that processes huge volume of multi-structured data in a highly reliable and scalable manner. Well-known examples of NoSQL are HBase and Cassandra.

Massively Parallel Processing (**MPP**) databases: MPPs are computational platforms that are able to process data at a very fast rate working basically on the concept of segmenting the data into chunks across different nodes in the cluster, and then processing the data in parallel. However, unlike Hadoop that operates at the disk level, MPP databases load the data into memory and operate on them using the collective memory of all the nodes in the cluster [1].

In the following sections we will see more examples of modern Big Data technologies.

DOI: 10.1201/9781003278177-12

BIG DATA TECHNOLOGY LANDSCAPE AND ANALYTICS PLATFORM

First let's discuss the essential components of a Big Data system, namely, processing frameworks and processing engines. A processing framework and engine together performs computation over the data in the system, either by reading from non-volatile storage or ingested directly into the system. More precisely, the *processing engine* is a component that is responsible for operating on data and the *processing framework* is a set of components designed to do the same. For instance, *Hadoop* is a processing framework in which the *MapReduce* is the default processing engine. However, in Big Data systems the engines and frameworks are often swapped out or used in tandem. Another example of the processing framework is *Apache Spark*, which uses the *MapReduce* as the processing engine.

Since we are exploring the processing of Big Data systems it is very important to discuss batch processing. *Batch processing* involves operating over a large, static dataset and returning the result later when the computation is complete. The characteristics of a batch processing datasets are (I) It represents a finite collection of data, (II) The data is almost always backed by some type of permanent storage for persistence, and (III) batch operations are often used for processing extremely large sets of data. It is important to remember that batch processing is well-suited for computations whenever the complete set of records is easily accessible. However, the tradeoff is the long processing time. Batch processing is not well suited for situations where the processing time is very significant [1].

The following components or layers work together to process batch data in Hadoop [1].

MapReduce: a parallel programming model for writing distributed applications to efficiently process large amounts of data on large clusters of commodity hardware in a reliable and fault-tolerant manner. The MapReduce program runs on Hadoop.

Hadoop Distributed File System (HDFS): provides a distributed file system that is designed to run on commodity hardware. It is similar to existing distributed file systems but is highly fault-tolerant and is designed to be deployed on low-cost hardware. It provides high throughput access to application data and supports applications having large datasets.

Hadoop Common: refers to Java libraries and utilities required by other Hadoop modules.

Hadoop YARN: a framework that supports job scheduling and cluster resource management.

Hbase: a column-based NoSQL database. It runs on top of HDFS and can handle any type of data. It allows for real-time processing and also supports random read/write operations to be performed in the data.

Pig: analyzes large datasets and overcomes the difficulty in writing the Map and Reduce functions. It consists of two components: Pig Latin and Pig Engine. Pig Latin is the scripting language similar to SQL. Pig Engine is the execution engine on which the Pig Latin runs.

Hive: a distributed data warehouse system. It allows for easy reading, writing, and managing files on HDFS. It has its own querying language for the purpose known as Hive Querying Language (HQL) which is very similar to SQL.

Sqoop: plays a major role in bringing data from Relational Databases into HDFS. The commands written in Sqoop internally convert it into MapReduce tasks that are executed over HDFS. It works with almost all relational databases such as MySQL, Postgres, SQLite, etc.

Zookeeper: used for the purpose of coordinating and synchronizing the nodes in a Hadoop cluster. It is an open source, distributed, and centralized service for maintaining configuration information, naming, providing distributed synchronization, and providing group services across the cluster.

In addition to the above-mentioned components, there are two additional components that are part of this ecosystem.

Kafka: Kafka a distributed message broker that sits between the applications generating data (Producers) and the applications consuming data (Consumers). Kafka is distributed and has in-built partitioning, replication, and fault-tolerance. It can handle streaming data and also allows businesses to analyze data in real-time.

Spark: an alternative framework to Hadoop built using the Scala language though it also supports varied applications written in Java, Python, R, etc. As an alternative to Hadoop, it provides in-memory processing which accounts for faster processing. In addition to batch processing, it can also handle real-time (stream) processing. Spark has a number of functions built within it to achieve specific Extraction, Transformation, and Loading (ETL) operations. The Spark Core is the main execution engine for Spark and the other APIs are built on top of it. For example, the Spark SQL API allows for querying structured data stored in data frames or Hive tables. Another API, the streaming API enables Spark to handle real-time data. It can easily integrate with a variety of data sources like Kafka, and Twitter.

Now that we have explored the different components of the Big Data technology landscape, we will discuss the data pipeline and its composition.

DATA PIPELINE ARCHITECTURE

A *data pipeline* is a series of tools for collecting the data, transforming it into insights, training models, delivering insights, and applying the model whenever and wherever the action needs to be taken to achieve the business goals. It essentially automates the ETL processes. A data pipeline architecture on the other hand is a system that captures, organizes, and routes data so that it can be used to gain insights. Raw data contains too many data points that may not be relevant. Data pipeline architecture organizes data events to make reporting, analysis, and using data easier. A customized combination of software technologies and protocols automate the management, visualization, transformation, and movement of data from multiple resources according to business goals [2].

At its core, any data pipeline will generally have five key components, namely, a data collection module, data ingestion module, data transformation module, computation/machine learning (ML) module, and data presentation module. At each stage, various applications can be used to perform the required operations to pass the data to the next stage. Sources for data collection includes a wide variety of sources such as cell phones, tablets, PCs, GPS sensors, etc. The data from the source has to be received and ingested through HTTP/MQTT endpoints either in the form of blobs or streams. Given the high volume and velocity of Big Data, a distributed messaging service such as Kafka can queue messages and broadcast messages based on topics. This simplifies the number of connections each client has to maintain with the various data sources. The data is then passed onto a data lake or to the staging area.

A *data lake* contains all the data in its raw form as it is received from the source. In the preparation stage, the data is transformed into structured data which can then be passed on to a data warehouse for secure storage and access. At this stage ETL is performed and structured/unstructured data is transformed into formats that can be consumed by applications and ML pipelines [3]. The second-to-last stage involves performing computations on the data to glean business intelligence and insights that add value to the business. These findings are then forwarded to the presentation stage where the results are delivered to the decision makers [4].

The Big Data pipeline is expansive with a wide array of data sources dealing with structured and unstructured data, being stored in the clouds or in distributed storage with multiple ETL pipelines feeding into various Business Intelligence (BI) and other applications.

LAMBDA ARCHITECTURE

Now that we have a broad view of the Big Data landscape, we can discuss a popular computing paradigm known as the Lambda architecture. The *Lambda architecture* is a hybrid approach that supports both batch- and stream-processing methods. This architecture has the capability to process massive quantities of data and can solve the problem of computing arbitrary functions. The Lambda architecture is composed of three layers, namely, batch, serving, and speed [5]. Let's briefly discuss each of the three layers.

In the *batch layer* the input data is fed continuously. This layer looks at all the data and eventually corrects the data. Many ETL tasks are performed in this layer and the data is stored on a data warehouse. This layer is built usually once or twice a day.

The *serving layer* has two very important functions namely: I) to manage the master dataset and II) to pre-compute the batch views. The outputs from the batch layer in the form of batch views are pushed to the speed layer in the form of near real-time views and these get forwarded for serving. This layer indexes the batch views so that they can be queried in an ad-hoc basis.

Finally, the *speed layer*, also known as the *stream layer*, handles the data that are not already in the batch view due to the latency of the batch layer. This layer deals only with recent data in order to provide a complete view of the data to the user by creating real-time views [5].

There are several benefits of this architecture including little server management as there is no need to install, maintain, or administer any software, automatically scaling applications by adjusting capacity, serverless applications with built-in availability and fault tolerance, and being able to adjust to real-time changing conditions. However, Lambda architecture is highly complex as there is a need for building two separate code bases for maintaining both the batch and the streaming layers. This arrangement makes the debugging process very complex [5].

Next, we will look at the lambda architecture–based data pipeline of two popular enterprises namely Twitter and Pinterest.

TWITTER AND PINTEREST'S DATA PIPELINE ARCHITECTURE

Twitter processes several hundred billion tweets per day generating a petabyte of daily data. The company's pipeline based on the lambda architecture is built upon a wide array of platforms including Hadoop, Kafka, BigQuery, PubSub, Vertica, and Manhattan. Twitter currently employs Scalding for batch processing, Heron for streaming data, and TimeSeries AggregatoR (TSAR) for both real-time and batch processing. Figure 12.1 provides a block diagram view of the data pipeline architecture of Twitter.

In Twitter's data pipeline, the data is fed simultaneously to the speed layer (called Heron) and to the batch layers. Within the batch layer, detailed ETL is performed, and the processed data is passed to the Master dataset. The speed layer can hold data until it is ready to be passed to the batch view layer. The batch component sources are Hadoop logs, such as client events, timeline events, and Tweet events, stored on HDFS [6]. Twitter employs several Scalding pipelines to preprocess the raw logs and ingests them into the Summing bird platform as offline sources, and within the Manhattan distributed storage systems. Within the stream processing, the real-time data from Kafka is stored in Nighthawk distributed cache. The TSAR Query service is then able to access both real-time and batch-stored data based on customer queries. To ensure fault tolerance, the data is replicated into three data centers [6].

Pinterest serves over 10 billion pageviews per month by heavily relying on their infrastructure consisting of Apache Kafka, Storm, Hadoop, HBase, and Redshift [7].

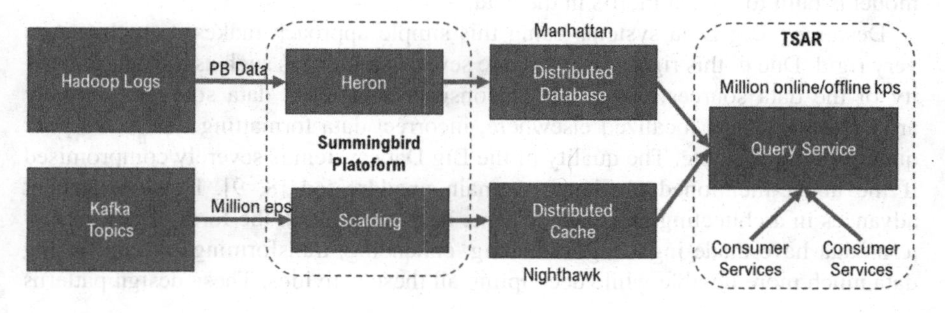

FIGURE 12.1 Block diagram of Twitter's architecture [6].

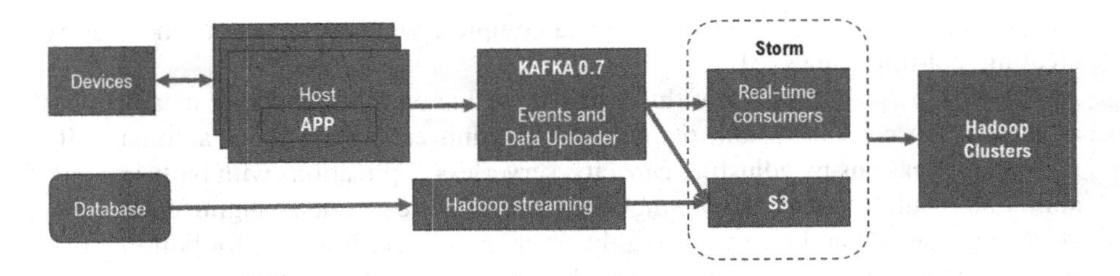

FIGURE 12.2 Block diagram of Pinterest's architecture [7].

At Pinterest, there are two primary categories of datasets: (I) online service logs and (II) database dumps. Service logs are continuous log streams generated by services across thousands of hosts whereas the database dumps are the logical database backups and are generated on hourly or daily basis.

Initially, Kafka is used as the central message transporter on the online service side. The app servers write log messages directly to Kafka. Once that is done, a data uploader on each Kafka uploads the Kafka log files to S3. A Hadoop streaming job pulls data from the database and writes the results into S3 (see Figure 12.2) [7].

In the next section we will discuss design strategies for building customized Big Data pipeline.

DESIGN STRATEGIES FOR BUILDING CUSTOMIZED BIG DATA PIPELINE

In the previous section we discussed the Big Data technology landscape and the Lambda architecture. In this section, the focus of discussion will be on the design strategies for building a customized Big Data pipeline. Depending upon a business use case, a Big Data pipeline can use a simple process, i.e., collect the data, load the data into a repository, and perform computation on the data. The data collected from diverse sources can be preprocessed and stored in the repositories that range from a simple file storage or a relational database to a complex data warehouse, data lake, or a data mart. The preprocessed data can then be channeled through ML algorithms for extracting features relevant to the use case. Finally, using the ML algorithms the model is built to extract motifs in the data.

Designing Big Data systems using this simple approach makes its architecture very rigid. Due to this rigidness, there are several challenges such as the heterogeneity of the data sources, errors, and inconsistencies in the data sources, untimely arrival of data, data localized elsewhere, incorrect data formatting types, and poor quality of the data, etc. The quality of the Big Data system is severely compromised if the above-mentioned challenges remain unaddressed [8, 9]. However, recent advances in architecting Big Data systems have emerged in the form of design patterns that have made ingesting, modeling, enhancing, transforming, and delivering data much more flexible while decoupling all these activities. These design patterns can also help architect Big Data systems with desired quality attributes such as performance, usability, maintainability, security, and reliability [9]. We'll start the discussion with the architectural design of the Big Data system and then followed by coverage of associated design patterns.

Yokoyama [10] described a widely used three-layer architecture mode for the Bid Data pipeline in enterprise applications called *Distinguish Business Logic from ML Model*. This architectural design aims at improving the operational stability of Big Data system (see Figure 12.3).

The striking observation here is the decoupling of the business logic from ML and a proposal for a three-layer architecture model. In his design the data has the potential to influence the business logic due to tight coupling with the ML module. This architectural design also identifies data dependency as one of the main technical debts in Big Data system.

Horizontally, this pattern decouples the Big Data system in to three layers namely the presentation layer, logic layer, and the data layer. The presentation layer focuses on designing user interfaces to facilitate user interaction and the data collection process. The logic layer consists of modules that implement the business logic, inference engines, and enables data preprocessing. Finally, the data layer focuses on designing databases, data marts and data lakes for storing the raw and processed data.

Vertically, the architectural design decouples the business logic specifics of Big Data system from its ML specifics. Business logic in the model encompasses the dashboard elements, and rules on how the reports should be interpreted and communicated. The inference engine is focused on data interpretation and pattern detection. Another key observation from Figure 12.3 is the decoupling of the inference engine

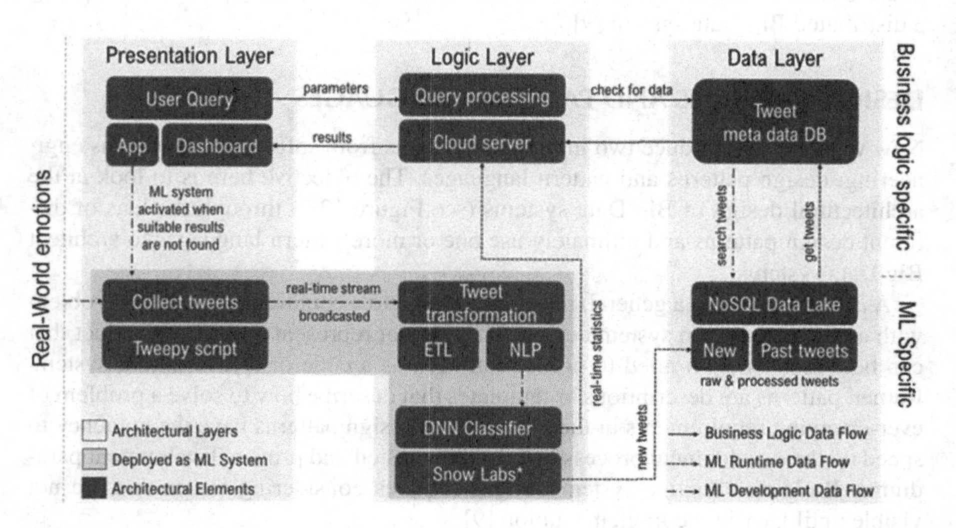

FIGURE 12.3 Architectural design of the Big Data pipeline [9, 10].

from the business logic. In such an architectural design, the ML subsystem can be easily updated and tested [9, 10].

This architectural design poses few challenges, however. For example, the ML algorithms derive their own set of functions, weights, and biases based only on the data that is available. This can be an inherent source of instability because the model's efficacy is contingent on the quality and the currency of the data that is being fed. Moreover, AI/ML, particularly in high value and high-profile systems implementations such as Twitter and Pinterest, are subject to adversarial attack (see Sidebar 1).

Therefore, in a tightly coupled system, the instability of the inference engine has the tendency to destabilize the business logic [9]. Other factors that can lower the efficacy of a Big Data system include changes in trends, data staleness, and lack of data on new items or queries. On the other hand, the proposed decoupling in Figure 12.3 makes the rollback process significantly easier as architects do not have to modify the business logic.

In summary, the proposed pattern segregates business logic from ML hierarchically by introducing four additional elements within the ML subsystem, namely, data collection, data processing, data lake, and inference engine. Another observation of the decoupled architecture in Figure 12.3 is the separation of data collection from the data storage. The data collection module continuously collects data from the sources and passes them onto the data processing module. This in-turn stores the cleaned data into the data lake, which ensures that the data collection is separate and independent from the user interface and the database.

The use of *Distinguish Business Logic from ML Model* architectural design strategy yields a structure with decoupled elements that can be deployed in a distributed manner. Separation of data collection, processing, and storage elements can help address some of the data quality issues and can be a good starting point for designing a distributed Big Data system [9].

DESIGN PATTERNS AND PATTERN LANGUAGES

Now we want to introduce two important concepts from software and systems engineering: design patterns and pattern languages. The objective here is to look at the architectural design of Big Data systems (see Figure 12.3) through the lens of different design patterns and ultimately use one or more pattern languages to architect Big Data systems.

A *design pattern* is a general, reusable solution to a commonly occurring problem with a given context in system design. They do not represent a finished product that can be directly transformed to production such as a code or a functioning system. Rather, patterns are descriptions or templates that describe how to solve a problem of ever-changing requirements in the system. The design patterns have the tendency to speed up the development process by providing tested and proven development paradigms. Building effective system designs requires considering issues that are not visible until later in the implementation [9].

Each time a system is designed from scratch without following any template, it can result in hidden subtle issues that take time for detection and repair. Such issues can cause major problems late in the system development cycle when the efforts and costs to repair are higher. Using design patterns, we can prevent such subtle issues and also improve the reusability of these systems.

PATTERN LANGUAGES are derived from a combination of different design patterns that together provide a solution to complex problems [11]. They are a way of expressing complex solutions derived from experience which can help others to gain a better understanding of the solution [9]. Here, it is important to understand that the *Distinguish Business Logic from ML Model* architectural design strategy is itself a design pattern. Let's look at some more examples of design patterns that are available for architecting Big Data Systems.

Table 12.1 highlights 17 different design patterns from which a subset of the design patterns can be composed together to form a pattern language for implementing fully functional distributed Big Data systems.

A handful of design patterns from Table 12.1 can be composed to create a pattern language for establishing a design strategy for Big Data System.

Let's take the example of an architectural design of the Big Data system shown in Figure 12.3 and derive a pattern language for this architecture. As mentioned before the *Distinguished Business Logic from ML Model* design pattern is clearly demonstrated by this architectural design of the Big Data system. In addition, the following design pattern is clearly demonstrated by this architectural design [9]:

1. *Workflow Pipeline*: The components within the Big Data system are containerized and are independently deployable. In addition to that the elements within the ML subsystem have been segregated to enable independent function.
2. *Real-Time Streaming*: The ML subsystem within this architectural design initiates real-time processing involving collection, transformation, and classification of data (tweets).
3. *Near-Real-Time*: The ML subsystem can perform near real-time processing of the unstructured data (tweets).
4. *Lightweight Stateless*: The ML subsystem is accessible through a web-based API.
5. *NoSQL*: Tweets are stored in non-relational format so that they can be quickly queried based on any of the predefined keys.
6. *Reproducibility*: Tweets are stored in multiple different partitions; one partition that stores all the raw tweets (and potentially other unstructured data such as images, audio and video) acting as a staging area for all raw data and another secondary partition that stores all processed and classified tweets.
7. *Explainable Predictions*: Explanations are provided for how and why the DNN Classifier makes its predictions to both understand the model and improve trust
8. *Checkpoints*: The steady state of the DNN model is preserved before and after each re-training phase. This ensures that the model employs the combination of weights and biases that minimizes the cost function for classification tasks.

TABLE 12.1
List of Design Patterns for Building a Reliable and Scalable Distributed Big Data System [9, 12–14]

Design Pattern	Layer	Specifics	Role of the Pattern	Importance
Rebalancing design pattern	Logic layer	ML specific	Focuses on strategies to deal with imbalanced datasets.	Required
Reproducibility design pattern	Logic layer	ML specific	Focuses on separating the input data from the features that encapsulate the preprocessing steps and include it into the model to ensure reproducibility.	Required
Checkpoints design pattern	Logic layer	ML specific	A checkpoint is a snapshot of the model's internal state so that training can be resumed from this state at any point in time. This design pattern is focused on providing resilience and fault-tolerance to architect scalable systems.	Required
Workflow pipeline design pattern	Presentation, logic, and data layer	Both Business logic and ML specifics	Isolates and containerizes the individual steps of a ML workflow into an organized workflow to ensure maintainability and scalability.	Required
Explainable design pattern	Logic layer	ML specific	A design pattern that can explain the model behavior aiding in diagnosing errors and in identifying biases in the model.	Optional
Multisource extractor	Presentation layer	Both business logic and ML specifics	An approach to ingest multiple data types from multiple data sources in an efficient manner. This pattern ensures high availability and distribution.	Required
Multi-destination design pattern	In between the presentation and the logic layer and in between the logic and the data layer	Business logic specific	Ingesting of raw data (after data collection) into HDFS and traditional data storage or other analytics platforms. This data pattern is highly scalable, flexible, fast, resilient to data failure, and cost-effective.	Optional

(Continued)

TABLE 12.1 (CONTINUED)
List of Design Patterns for Building a Reliable and Scalable Distributed Big Data System [9, 12–14]

Design Pattern	Layer	Specifics	Role of the Pattern	Importance
Protocol converter	Logic and data layer	Both Business logic and ML specific	A mediatory approach to provide an abstraction for the incoming data from various systems. This design pattern provides an efficient way to ingest a variety of unstructured data from multiple data sources and different protocols.	Optional
Just-In-Time design pattern	Logic layer	Both business logic and ML specific	Used in situations where data needs to be preloaded in the data stores before transformation and preprocessing can happen. This pattern runs independently performing cleaning, validating, correlating, transforming, and storing resultant data in the data store.	Optional
Real-time streaming pattern	Presentation and logic layer	ML specific	Facilitates continuous and real-time processing of the unstructured data. This pattern can minimize latency, help build scalable systems, facilitating the parsing of real-time information, etc.	Required
Workload balancing patterns	Logic layer	ML specific mostly	A set of 11 design patterns constituted together as a workload balancing pattern. Primarily, these patterns help to address data workload challenges associated with different domains and business cases efficiently.	Required
Façade pattern	Data layer	Business logic specific	Facilitates communication between the different data sources in the enterprise and the business intelligence tools.	Optional
NoSQL pattern	Data layer	Business logic specific	Offers NoSQL alternatives in place of traditional RDBMS to facilitate the rapid access and querying of Big Data.	Optional

(*Continued*)

TABLE 12.1 (CONTINUED)

List of Design Patterns for Building a Reliable and Scalable Distributed Big Data System [9, 12–14]

Design Pattern	Layer	Specifics	Role of the Pattern	Importance
Connector pattern	Between data and logic layer	Both business logic and ML specific	Provides a developer API and SQL like query language to access the data and to gain significantly reduced development time.	Optional
Lightweight stateless pattern	Between data and logic layer	Both business logic and ML specific	Provides data access through web services, and it is independent of platform or language implementations.	Optional
Near-real-time pattern	Between data and logic layer	ML specific	Implements solutions for near real-time data access.	Optional
Stage transform pattern	Between data and logic layer	Business logic specific	Reduces the data scanned and fetched based upon business needs.	Optional

SUMMARY

In summary, we have discussed the Big Data technology landscape and the analytics platform, introduced the concept of Lambda architecture, and have seen how patterns and pattern languages can be used to architect customized Big Data pipeline that exhibits various quality attributes such as reliability, maintainability, usability, and enhanced performance, and security.

SIDEBAR 1 ADVERSARIAL MACHINE LEARNING TECHNIQUES

A significant concern for designers of machine learning algorithms is adversarial attacks. Here, malicious individuals access the training data during the training or testing phases, modifying model predictions, creating bias, and ruining the application for the purpose of espionage, sabotage, or fraud. Attackers can also inject corrupted samples after deployment, altering the effectiveness of the algorithm as it learns incorrectly.

For example, consider an AI system that was designed to recognize people in images. In a certain type of attack called data poisoning, introducing adversarial samples that include a purposefully designed t-shirt pattern can fool the AI into not identifying people. In another example, specially designed eyeglass frames were able to deceive a state-of-the-art facial recognition system, enabling the wearers to "disappear" or to appear as a different person, even certain celebrities.

While financial losses can be incurred in an attacked ML system, if the system is used in some sort of critical application, for example, disease diagnosis, the results could be catastrophic and cost human lives. Therefore, it is important that the ML system designer prepare defenses against such attacks.

The U.S. National Standards Institute (NIST) created a taxonomy and classification of adversarial machine learning techniques. They also describe a number of defenses against adversarial ML [NIST]. Defenses during training and testing include data encryption, data sanitization and robust statistics, the latter approach involving the use of constraints and regularization techniques to reduce potential distortions of the learning model caused by poisoned data. Defenses that can be used against training attacks, however, often can incur performance overhead as well as have a detrimental effect on model accuracy. Defenses against attacks that corrupt learning in ML algorithms require purposeful and robust design. Research is ongoing on the best approaches for these types of designs.

[NIST] Elham Tabassi, Kevin J. Burns, Michael Hadjimichael, Andres D. Molina-Markham and Julian T. Sexton, "Draft NISTIR 8269," "A Taxonomy and Terminology of Adversarial Machine Learning," National Institute of Standards and Technology, October, 2019, https://nvlpubs.nist.gov/nistpubs/ir/2019/NIST.IR.8269-draft.pdf.

EXERCISE

1. Discuss the merits and demerits of the Hadoop computing framework. Also compare its processing capabilities against the spark computing framework.

2. Using suitable examples discuss the merits and demerits of the lambda architecture.

3. Discuss the role of all the required design patterns for building a reliable and scalable distributed Big Data pipeline.

4. The design pattern that isolates and containerizes the individual steps of a Machine Learning workflow into an organized workflow to ensure maintainability and scalability is
   A. Reproducibility design pattern
   B. Explainable design pattern
   C. Workflow pipeline design pattern
   D. Rebalancing design pattern
   E. Multi-destination design pattern

5. A ________________ contains all the data in its raw form as it is received from the source.
   A. Data Warehouse
   B. Data lake
   C. Data pipeline
   D. ML pipelines
   E. RDBMS

6. ____________ is used for the purpose of coordinating and synchronizing the nodes in a Hadoop cluster.
   A. Hadoop YARN
   B. Pig
   C. Hadoop
   D. Zookeeper
   E. Kafka

REFERENCES

1. Gupta, S., Saxena, S. (2016). *Real-Time Big Data Analytics*, Packt Publishing Ltd., ISBN 978-1-78439-140-9.
2. Snaplogic. "Data Pipeline Architecture", retrieved from https://www.snaplogic.com/glossary/data-pipeline-architecture, retrieved on July 19, 2022.
3. John Snow Labs Inc. (2021). "Emotion Detection Classifier", retrieved from https://nlp.johnsnowlabs.com/2021/01/09/classifierdl_use_emotion_en.html.
4. Gupta, S.C. "Architecture for High-Throughput Low-Latency Big Data Pipeline on Cloud", retrieved from https://towardsdatascience.com/scalable-efficient-big-data-analytics-machine-learning-pipeline-architecture-on-cloud-4d59efc092b5, retrieved on July 19, 2022.
5. Anonymous. "Lambda Architecture", retrieved from https://databricks.com/glossary/lambda-architecture, retrieved on July 19, 2022.
6. Malife, C. "Processing Billions of Events in Real Time at Twitter", retrieved from https://blog.twitter.com/engineering/en_us/topics/infrastructure/2021/processing-billions-of-events-in-real-time-at-twitter, retrieved on July 19, 2022.
7. Yang, Y. "Scalable and Reliable Data Ingestion at Pinterest", retrieved from https://medium.com/pinterest-engineering/scalable-and-reliable-data-ingestion-at-pinterest-b921c2ee8754, retrieved on July 19, 2022.
8. Lakshmanan, V., Robinson, S., Munn, M. (2020). *Machine Learning Design Patterns*. O'Reilly Media, Inc.
9. Srinivasan, S. M., Mahbub, S., Sangwan, R. S., Badr, Y., Mukherjee, P. (2022). "Pattern Language for Designing Distributed AI Systems", published in the 2022 INFORMS Conf. on Service Science, China.
10. Yokoyama, H. (2019). Machine Learning System Architectural Pattern for Improving Operational Stability. *IEEE International Conference On Software Architecture Companion*, 267–274. doi:10.1109/ICSA-C.2019.00055.
11. Buschmann, F., Henney, K., Schmidt, D. (2007). *Pattern Oriented Software Architecture: On Patterns and Pattern Languages*. Wiley Software Patterns Series, John Wiley & Sons, Inc.
12. Washizaki, H. et al., (2022). Software-Engineering Design Patterns for Machine Learning Applications. *Computer*, 55(3), 30–39. 10.1109/MC.2021.3137227.
13. Raj, P., Raman, A., Subramanian, H. (2017). *Architectural Patterns*, Packt Publishing Ltd., ISBN 978-1-78728-749-5.
14. International Organization for Standardization. (2011, March). "Systems and Software Engineering — Systems and Software Quality Requirements and Evaluation (SQuaRE) — System and Software Quality Models". retrieved from iso.org: https://www.iso.org/obp/ui/#iso:std:iso-iec:25010:ed-1:v1:en, retrieved on April 24, 2022.

Glossary of Terms

activation function in ANN, the activation function of a node defines the output of that node given an input or set of inputs.

AdaBoost an ensemble learning method that uses an iterative approach to learn from the mistakes of weak classifiers and turn them into strong ones.

agglomerative hierarchical clustering a bottom-up approach in which each observation starts as its own cluster, and pairs of clusters are merged as one moves up the hierarchy.

analysis of variance (ANOVA) an analysis tool used in statistics that splits an observed aggregate variability found inside a dataset into two parts: systematic factors and random factors.

Apache Spark a multilanguage engine for executing data engineering, data science, and machine learning on single-node machines or clusters.

apriori algorithm refers to an algorithm that is used in mining frequent item sets and relevant association rules.

artificial intelligence (AI) refers to systems or machines that mimic human intelligence to perform tasks and can iteratively improve themselves based on the information they collect.

artificial neural network (ANN) one of a subset of machine learning that is at the heart of deep learning algorithms. The name and structure are inspired by the human brain, mimicking the way biological neurons signal to one another.

association rule mining a machine learning model that identifies frequent if-then associations, also known as association rules.

autocorrelation a statistical measure that in data analytics represents the degree of similarity between a given time series and a lagged version of itself over successive time intervals.

backward propagation the process of moving from the output layer (right) to the input layer (left) in a neural network.

bag of words (BOW) a model that is a simplifying representation used in natural language processing and information retrieval where text is represented by the words used, and the multiplicity of the time each word is used. Grammar and word order are disregarded.

bagging a homogeneous weak learners' model that learns from each other independently in parallel and combines them for determining the model average.

batch layer in the lambda architecture a layer that looks at all the data and eventually corrects the data. Many ETL tasks are performed in this layer

batch processing when a computer processes a number of tasks that it has collected in a group.

batch size defined as the number of training examples in one forward/backward pass. The higher the batch size, the more the memory space consumed.

Bayesian classifier a probabilistic model where the classification is a latent variable that is probabilistically related to the observed variables.

bias a phenomenon that occurs when an algorithm produces results that are systemically prejudiced due to erroneous assumptions in the machine learning process.

Big Data refers to extremely large datasets that may be analyzed computationally to discover patterns, trends, and associations, especially relating to human behavior and interactions.

bimodal distribution when a set of data is distributed in two clusters.

boosting a model in which the weak learners learn sequentially and adaptively to improve the model predictions of a learning algorithm.

Box–Cox transformation a transformation of non-normal dependent variables into a normal shape.

boxplot a method for graphically demonstrating the locality, spread, and skewness groups of numerical data through their quartiles. It is also known as **box plot**.

business intelligence (BI) a technology-driven process for analyzing data and delivering actionable information that helps executives, managers, and workers make informed business decisions.

central limit theorem states that if you have a population with mean μ and standard deviation σ and take sufficiently large random samples from the population with replacement, then the distribution of the sample means will be approximately normally distributed.

chi-square a statistical test used to examine the differences between categorical variables from a random sample in order to judge goodness of fit between expected and observed results.

clustering an unsupervised machine learning task that involves automatically discovering natural grouping in data.

collaborative filtering a technique that can filter out items that a user might like on the basis of reactions by similar users. It works by searching a large group of people and finding a smaller set of users with tastes similar to a particular user.

complete linkage one of the agglomerative hierarchical clustering techniques. At the beginning of the process, each element is in a cluster of its own. The clusters are then sequentially combined into larger clusters until all elements end up being in the same cluster.

conditional probability refers to the chances that some outcome occurs given that another event has also occurred.

content-based filtering a type of recommender system that attempts to guess what a user may like based on that user's activity. Content-based filtering makes recommendations by using keywords and attributes assigned to objects in a database and matching them to a user profile.

convolution layer a layer where filters are applied to the original image, or to other feature maps in a deep CNN. Convolution layers are where most of the user-specified parameters are in the network.

convolution a mathematical operation that allows the merging of two sets of information. In the case of CNN, convolution is applied to the input data to filter the information and produce a feature map.

convolutional neural network (CNN) a Deep Learning algorithm, which can take in an input image, assign importance (learnable weights and biases) to various aspects/objects in the image, and be able to differentiate one from the other.

cophenetic correlation coefficient a measure of how accurately and reliably a dendrogram preserves the pairwise distance between the original unmodeled data points.

corpus a large and unstructured set of texts used to do statistical analysis and hypothesis testing, checking occurrences or validating linguistic rules within a specific language.

correlation coefficient a statistical measure of the strength of the relationship between the relative movements of two variables.

cosine similarity a measure of the similarity between two vectors of an inner product space. It is measured by the cosine of the angle between two vectors and determines whether two vectors are pointing in roughly the same direction.

covariance a measure of the joint variability of two random variables.

cross-validation technique a technique for evaluating machine learning models by training them on subsets of the available input data and evaluating them on the complementary subset of the data.

data acquisition phase a stage of analytics in which the metadata is generated to describe what data needs to be recorded and measured.

data cleaning the process of fixing incorrect, incomplete, duplicate, or otherwise erroneous data in a dataset. Also referred to as data cleaning or data scrubbing.

data deduplication in computing, a technique for eliminating duplicate copies of repeating data.

data discretization a process of converting a large number of data values into smaller ones.

data integration, aggregation, and representation phase a data analysis phase that hides the heterogeneity of the data and makes it available in the required format for analysis and modeling.

data integration the process of combining data from multiple source systems to create unified sets of information for both operational and analytical uses.

data pipeline set of tools and processes used to automate the movement and transformation of data between a source system and a target repository.

data provenance refers to the record trail that accounts for the origin of a datum together with an explanation of how and why it got to the present place.

data reduction the process to obtain a reduced representation of the dataset that is much smaller in volume but yet produces the same analytical results.

data transformation the process of converting data from one type to another.

data warehouse a central repository of information that is collected from transactional systems, relational databases, and other sources, on a regular basis and is analyzed to make more informed decisions.

dataframe a data structure that organizes data into a 2-dimensional table of rows and columns, much like a spreadsheet.

date lake a centralized repository designed to store, process, and secure large amounts of structured, semi-structured, and unstructured data.

deep belief network (DBN) a generative graphical model composed of multiple layers of latent variables, with connections between the layers but not between units within each layer.

deep learning (DL) a subset of machine learning that is designed to function like the human brain using artificial neural networks.

deep neural networks (DNN) a neural network with some level of complexity, usually at least two layers in addition to the input and output layer.

dendrogram a diagram that shows the hierarchical relationship between objects.

design pattern a general, reusable solution to a commonly occurring problem with a given context in system design. There are many design patterns. None represents a finished product. Each is a template that describes how to solve a problem of ever-changing requirements in the system.

determinant in mathematics, a scalar value that is a function of the entries of a square matrix. It allows characterizing some properties of the matrix and the linear map represented by the matrix.

dirty with respect to data, refers to corrupt, inconsistent, or uncertain data.

divisive hierarchical clustering a top-down approach in which all observations start in one cluster, and splits are performed recursively as one moves down the hierarchy.

doubledecker plot visualizes the dependence of one categorical (typically binary) variable on further categorical variables.

dropout a technique where randomly selected neurons are ignored during the training stage. They are "dropped out" randomly.

eigenvalue one of a special set of scalars associated with a linear system of equations (i.e., a matrix equation) that are sometimes also known as characteristic roots, characteristic values, or latent roots.

ElasticNet regressors (L1/L2 norm) a form of linear regression that uses the penalties from both the lasso and ridge techniques to regularize regression models.

ensemble method a technique that creates multiple models and then combines them to produce improved results. Ensemble methods usually produce more accurate solutions than a single model would.

entropy in information theory is the average amount of information conveyed by an event, when considering all possible outcomes.

epoch a hyperparameter that defines the number of times that the learning algorithm will work through the entire training dataset.

error term a value which represents how observed data differs from actual population data.

ETL acronym for "extract, transform, and load." This process is used by data engineers to extract data from different sources, transform the data into a usable and trusted resource, and load that data into the systems where end-users can access and perform downstream analysis to solve business problems.

Euclidean distance the distance between two points in Euclidean space is the length of a line segment between the two points.

experiment any procedure that can be infinitely repeated and has a well-defined set of possible outcomes, known as the sample space. Also known as trial.

explainability In AI and ML, indicates that the systems designers can rationalize the system decision making, characterize their strengths and weaknesses, and convey an understanding of how the system will behave in the future.

feed forward neural network (FNN) an ANN where connections between the nodes do not form a cycle. In this network, the information moves in only one direction—forward—from the input nodes, through the hidden nodes (if any), and to the output nodes.

forward propagation the way to move from the Input layer (left) to the Output layer (right) in the neural network.

Gini impurity used to predict the likelihood that a randomly selected example would be incorrectly classified by a specific node. Pronounced "genie."

gradient descent an optimization algorithm which is commonly used to train machine learning models and neural networks.

Hadoop Distributed File System (HDFS) the primary storage system used by Hadoop applications. This open-source framework works by rapidly transferring data between nodes. It's often used by companies who need to handle and store Big Data.

Hadoop a collection of open-source software utilities that facilitates using a network of many computers in a distributed environment to solve problems involving massive amounts of data and computation. Runs on the Apache HTTP open-source server.

heterogeneity a challenge associated with data representing the same entity in different sources lacking a consistent format, being incomplete and erroneous.

hidden layer an intermediate layer between the input and the output layer containing neurons. It is a place where all the computation is done.

hierarchical clustering a method of cluster analysis which seeks to build a hierarchy of clusters.

homoscedastic for variances, is an assumption of equal or similar variances in different groups being compared. This is an important assumption of parametric statistical tests. Also known as homogeneity of variances.

human intervention and collaboration a process that mitigates one of the challenges to the automation of the data analysis pipeline where humans and machines work together to identify patterns that are only detectable by humans and missed by machines.

hybrid recommendation system (HRS) a special type of recommendation system which can be considered as the combination of the content and collaborative filtering method.

identity matrix a square matrix whose diagonal entries are all equal to one and whose off-diagonal entries are all equal to zero.

information extraction and cleaning phase in data analytics the process responsible for converting the collected data into a format that is ready for analysis.

information gain measures the reduction in entropy or surprise by splitting a dataset according to a given value of a random variable.

information technology the study or use of computers and telecommunications for storing, retrieving, and sending information.

interpretation phase provides the means for the decision makers to interpret the results of the analysis and make the Big Data more actionable.

item-based collaborative filtering (IBCF) a type of recommendation system that is based on the similarity between items calculated using the rating users have already given to the items.

kurtosis a measure of the combined weight of a distribution's tails relative to the center of the distribution.

lambda architecture a data-processing architecture designed to handle massive quantities of data by taking advantage of both batch and stream-processing methods.

lasso regression (L1 norm) a method usually used in machine learning for the selection of the subset of variables. It provides greater prediction accuracy as compared to other regression models. Also called Penalized regression method.

learning rate a hyperparameter that controls how much to change the model in response to the estimated error each time the model weights are updated.

likelihood the state or fact of some event being likely.

long short-term memory (LSTM) a type of RNN capable of learning order dependence in sequence prediction problems. This is a behavior required in complex problem domains like machine translation, speech recognition, and more.

machine learning (ML) a subset of AI and is the science of training devices or software to perform a task and improve its capabilities by giving it data so it can "learn" over time.

Manhattan distance a distance metric between two points, and it is computed as the sum of the lengths of the projections of the line segment between the points onto the coordinate axes.

MapReduce a programming model and an associated implementation for processing and generating big datasets with a parallel, distributed algorithm on a cluster. A MapReduce program is composed of a map procedure, which performs filtering and sorting, and a reduce method, which performs a summary operation.

matrix In mathematics, a rectangular array or table of numbers, symbols, or expressions, arranged in rows and columns, which is used to represent a mathematical object or a property of such an object.

mean absolute error (MAE) with respect to a test set of a model is the mean of the absolute values of the individual prediction errors over all the instances in the test set.

mean squared error (MSE) measures how close a regression line is to a set of data points.

missing at random (MAR) when the probability of a record having a missing value for an attribute could depend on the observed data, but not on the value of the missing data itself. Data which is incomplete only due to structural reasons are MAR.

missing completely at random (MCAR) when the probability of a record having a missing value for an attribute does not depend on either the observed data or the missing data.

multicollinearity a statistical concept where several independent variables in a model are correlated.

multilayer perceptron (MLP) a fully connected class of feed-forward ANN.

multiple regression analysis a statistical technique that can be used to analyze the relationship between a single dependent variable and several independent variables.

multivariate logistic regression (MLogR) analysis performed to predict the relationships between dependent and independent variables. It calculates the probability of something happening depending on multiple sets of variables. This is a common classification algorithm used in data science and machine learning.

natural language processing (NLP) a subfield of linguistics; computer science; and AI concerned with programming computers to process and analyze large amounts of natural language data.

neuron a connection point in an artificial neural network. Artificial neural networks, like the human body's biological neural network, have a layered architecture and each network node has the capability to process input and forward output to other nodes in the network.

***n*-gram** in computational linguistics a contiguous sequence of *n* items from a given sample of text or speech.

normal distribution an arrangement of a dataset in which most values cluster in the middle of the range and the rest taper off symmetrically toward both the extremes.

not missing at random (NMAR) in a set of data when the probability of a record having a missing value for an attribute could depend on the value of the attribute. Missing data mechanism that is considered as NMAR is non-ignorable.

orthogonal matrix a square matrix with real numbers or elements where its transpose is equal to its inverse matrix. When the product of a square matrix and its transpose gives an identity matrix, then the square matrix is known as an orthogonal matrix.

outcome in probability theory, a possible result of an experiment or trial. Each possible outcome of a particular experiment is unique, and different outcomes are mutually exclusive (only one outcome will occur on each trial of the experiment).

outlier an observation that lies an abnormal distance from other values in a random sample from a population.

overfitting a modeling error in statistics that occurs when a function is too closely aligned to a limited set of data points. As a result, the model is useful in reference only to its initial dataset, and not to any other datasets.

padding a term relevant to CNN as it refers to the number of pixels added to an image when it is being processed by the kernel of a CNN.

part-of-speech (POS) tagging a popular NLP process which refers to categorizing words in a text (corpus) in correspondence with a particular part of speech, depending on the definition of the word and its context.

pattern language a collection of different design patterns that together provide a solution to complex problems.

Pearson correlation coefficient a single number that measures both the strength and the direction of the linear relationship between two continuous variables.

perceptron a neural network unit that does certain computations to detect features or business intelligence in the input data. It is a function that maps its input which is multiplied by the learned weight coefficient and generates an output value.

pooling layer In CNN a pooling layer is used to reduce the dimensions of the feature maps. Thus, it reduces the number of parameters to learn, and the amount of computation performed in the network.

posterior probability in Bayesian statistics, is the revised or updated probability of an event occurring after taking into consideration new information.

principal component analysis (PCA) the process of computing the principal components and using them to perform a change of basis on the data, sometimes using only the first few principal components and ignoring the rest.

priori probability refers to the likelihood of an event occurring when there is a finite number of outcomes, and each is equally likely to occur. The outcomes in a priori probability are not influenced by the prior outcome.

privacy In Big Data, a major concern as there is no established protocol that allows the sharing of private data while limiting the disclosure and ensuring sufficient data utility.

probability distribution a mathematical function that describes the probability of different possible values of a variable. They are often depicted using graphs or probability tables.

processing engine in the data pipeline the entity responsible for processing data, usually retrieved from storage devices, based on a predefined logic, in order to produce a result.

processing framework allows users to process data in a Hadoop cluster using the low-level API.

proportion of variance a generic term to mean a part of variance as a whole.

query processing, data modeling, and analysis phase The phase that deals with building tools and techniques for effective large-scale analysis of data in a completely automated manner.

random forest an ensemble learning method for classification, regression, and other tasks that operates by constructing a multitude of decision trees at the training time. Also known as random decision forest.

random variable a variable whose value is unknown or a function that assigns values to each of an experiment's outcomes. A random variable can be either discrete (having specific values) or continuous (any value in a continuous range).

recommender engine software that analyzes available data to make suggestions for something that a website user might be interested in.

recommender system a subclass of Information Filtering Systems that seeks to predict the rating or the preference a user might give to an item.

recurrent neural network (RNN) a class of ANN where connections between nodes can create a cycle, allowing output from some nodes to affect subsequent input to the same nodes. This allows it to exhibit temporal dynamic behavior.

regularization refers to techniques that are used to calibrate machine learning models in order to minimize the adjusted loss function and prevent overfitting or underfitting.

residual In linear regression, defined as the difference between an observed value of a response variable and the value of the response variable predicted from the regression line.

ridge regression (L2 norm) a model tuning method that is used to analyze any data that suffers from multicollinearity. This method performs L2 regularization. When the issue of multicollinearity occurs, least-squares are unbiased, and variances are large, this results in predicted values being far away from the actual values.

root mean square error (RMSE) the standard deviation of the residuals.

sample space a random experiment is the collection of all possible outcomes.

scale a challenge associated with the management of large and rapidly increasing volume of data given that the tools, techniques, and algorithms that process them have severe limitations.

scree plot In multivariate statistics, a line plot of the eigenvalues of factors or principal components in an analysis. The scree plot is used to determine the number of principal components to keep in a PCA.

semi-structured data information that does not reside in a relational database but has some organizational properties that make it easier to analyze.

sentiment classification the automated process of identifying opinions in text and labeling them as positive, negative, or neutral, based on the emotions customers express within them.

serving layer in the lambda architecture the layer that manages the master dataset and pre-computes the batch views. This layer indexes the batch views so that they can be queried in an *ad hoc* basis.

singular value decomposition (SVD) In linear algebra a factorization of a matrix into three matrices. SD has some interesting algebraic properties and conveys important geometrical and theoretical insights about linear transformations.

speed layer in the lambda architecture the layer that is used to handle the data that are not already in the batch view due to the latency of the batch layer.

statistics the science of collecting, analyzing, presenting, and interpreting data.

stemming In NLP the process of reducing a word to its word stem that affixes to suffixes and prefixes or to the roots of words known as a lemma.

stopword In NLP a word which is filtered out before or after processing the text data because it is insignificant.

stream processing a data management technique that involves ingesting a continuous data stream to quickly analyze, filter, transform, or enhance the data in real-time.

stride In CNN the number of pixels shifts over the input matrix.

structurally missing data that is missing for a logical reason. It is data that is missing because it should not exist.

structured data data whose elements are effectively organized into a formatted repository that is typically a database.

supervised learning a subcategory of machine learning and AI. It uses labeled datasets to train algorithms that can then classify data or predict outcomes accurately.

timeliness the challenge of analyzing data in a timely manner. Often analyzing large datasets takes a longer time, which is a challenge because over the period of time the value of the data diminishes for a decision maker.

tokenization in NLP the process of splitting a phrase, sentence, paragraph, or an entire text document into smaller units, such as individual words or terms.

transformation in data analysis the replacement of a variable by a function of that variable.

underfitting a situation when a data model is unable to capture the relationship between the input and output variables accurately, generating a high error rate on both the training set and unseen data.

unstructured data data which is not organized in a predefined manner or does not have a predefined data model; thus, it is not a good fit for a mainstream relational database.

unsupervised learning refers to the use of algorithms to identify patterns in datasets containing data points that are neither classified nor labeled.

user-based collaborative filtering (UBCF) a technique used to predict the items that a user might like on the basis of ratings given to that item by the other users who have similar taste with that of the target user.

value a dimension of Big Data that usually refers to the insight discovery and pattern recognition that results in more effective operations, stronger customer relationships and other clear and quantifiable business benefits.

vanishing gradient problem when there are more layers in the network, the value of the product of the derivative decreases until at some point the partial derivative of the loss function approaches a value close to zero, and the partial derivative vanishes.

variance inflation factor (VIF) a measure of the amount of multicollinearity in a set of multiple regression variables.

variety a dimension of Big Data that refers to the diversity and range of different data types, including unstructured data, semi-structured data, and raw data.

vector a quantity having direction as well as magnitude, especially as determining the position of one point in space relative to another.

velocity a dimension of Big Data that refers to the speed at which an organization receives, stores, and manages data.

veracity a dimension of Big Data that refers to the accuracy of data and information assets, which often determines executive-level confidence.

volume a dimension of Big Data that refers to the size and amounts of Big Data that companies manage and analyze.

weak learner refers to simple models that do only slightly better than random chance.

white noise refers to residuals that are random and come from a single distribution.

XGBoost a scalable, distributed gradient-boosted decision tree machine learning library that provides parallel tree boosting and is the leading machine learning library for regression, classification, and ranking problems.

yet another resource negotiator (YARN) allows the data stored in HDFS to be processed and run by various data processing engines such as batch processing, stream processing, etc. Runs on the Apache HTTP open-source server.

zookeeper a centralized service for maintaining configuration information, naming, providing distributed synchronization, and providing group services. All of these kinds of services are used in some form or another by distributed applications. Runs on the Apache HTTP open-source server.

Index

Page numbers in **bold** indicate tables, page numbers in *italic* indicate figures and page numbers followed by n indicate notes.

U

V

W

X

Y

Z

Printed in the United States
by Baker & Taylor Publisher Services

Printed in the United States
by Baker & Taylor Publisher Services